Drought - Impacts and Management

*Edited by Murat Eyvaz,
Ahmed Albahnasawi, Mesut Tekbaş
and Ercan Gürbulak*

Published in London, United Kingdom

Drought - Impacts and Management
http://dx.doi.org/10.5772/intechopen.97906
Edited by Murat Eyvaz, Ahmed Albahnasawi, Mesut Tekbaş and Ercan Gürbulak

Contributors
Kousik Atta, Aditya Pratap Singh, Saju Adhikary, Subhasis Mondal, Sujaya Dewanjee, Viktor Lyakh, Anatoliy Soroka, Md. Shafiqul Islam, Md. Nazrul Islam Mukul, Godwill Makunde, Nation Chikumba, Walter Svinurai, Xavier Mhike, Pooja Dhansu, Arun Kumar Raja, Krishnapriya Vengavasi, Ravinder Kumar, Ashwani Kumar, Naresh Kumar, Anita Mann, Shashi Kant Pandey, Adhini S. Pazhany, Kaltoum Belhassan, Kassa Abera, Admasu Gebeyehu, Florin Nenciu, Eugen Popescu, Valentin Nicolae Vladut, Chelea Matchawe, Patrice Bonny, Germaine Yandang, Huguette Cecile Yangoua Mafo, Bonglaisin J. Nsawir, R. Rintis Hadiani, Bambang Suharto, Agus Suharyanto, Suhardjono

First published in London, United Kingdom, 2022 by IntechOpen
IntechOpen is the global imprint of INTECHOPEN LIMITED, registered in England and Wales, registration number: 11086078, 5 Princes Gate Court, London, SW7 2QJ, United Kingdom

British Library Cataloguing-in-Publication Data
A catalogue record for this book is available from the British Library

Additional hard and PDF copies can be obtained from orders@intechopen.com

Drought - Impacts and Management
Edited by Murat Eyvaz, Ahmed Albahnasawi, Mesut Tekbaş and Ercan Gürbulak
p. cm.
Print ISBN 978-1-80355-543-0
Online ISBN 978-1-80355-544-7
eBook (PDF) ISBN 978-1-80355-545-4

Meet the editors

Dr. Murat Eyvaz is an associate professor in the Environmental Engineering Department, Gebze Technical University, Turkey. His research interests include applications in water and wastewater treatment facilities, electrochemical treatment processes, filtration systems at the lab, pilot-scale membrane processes (forward osmosis, reverse osmosis, membrane bioreactors), membrane manufacturing methods (polymeric membranes, nanofiber membranes, electrospinning), spectrophotometric analyses (UV, atomic absorption spectrophotometry), and chromatographic analyses (gas chromatography, high-pressure liquid chromatography). He has co-authored many journal articles and conference papers and has taken part in many national projects. He serves as an editor and reviewer for numerous journals. He holds four patents on wastewater treatment systems.

Dr. Ahmed Albahnasawi is a post-doctorate fellow in the Environmental Engineering Department, Gebze Technical University, Turkey. His graduate work focused on the investigation of the treatability of the sequential anoxic-aerobic batch reactors followed by ceramic membrane for textile wastewater treatment. Dr. Albahnasawi has published three journal papers based on his Ph.D. research. He has participated in three international conferences. His research interests include the application and design of a microbial fuel cell integrated with Fenton oxidation for industrial wastewater treatment/solid waste management and monitoring of organic micropollutants by both chromatographic and spectrophotometric analyses.

Dr. Ercan Gürbulak is a research associate in the Environmental Engineering Department, Gebze Technical University, Turkey. He received his bachelor's degree in Environmental Engineering at Marmara University, Turkey in 2005. He obtained an MSc and Ph.D. from Gebze Technical University in 2008 and 2019, respectively. His research interests include the application and design of hydrothermal processes for industrial wastewater treatment/solid waste management and monitoring of organic micropollutants by both chromatographic and spectrophotometric analyses.

Dr. Mesut Tekbaş is a researcher/lecturer in the Environmental Engineering Department, Gebze Technical University, Turkey. He received his bachelor's degree in Environmental Engineering at Ondokuz Mayıs University, Turkey in 2003. He obtained an MSc and Ph.D. from Gebze Technical University in 2007 and 2019, respectively. His research interests include the application and design of supercritical water oxidation processes for wastewater treatment/solid waste management and electrochemical analyses.

Contents

Preface

Drought is generally defined as a prolonged period of dry weather caused by a lack of precipitation that results in severe water shortages for some activities, populations, or ecological systems. The deterioration of the balance between precipitation and evaporation and the long duration of this situation can also be considered a drought. Drought types, such as meteorological, agricultural, hydrological, and socioeconomic droughts, adversely affect many environmental components, including soil processes, vegetation growth, wildlife, water quality, and aquatic ecosystems. While drought has unfavorable impacts on both surface and groundwater resources, hydrological regimes can also be affected by it, changing the chemistry of surface waters and the runoff pathway, which can negatively influence water quality.

While drought appears as one of the main consequences of changes in ecosystem and climate, the consumption of water used for domestic, industrial, and agricultural purposes has increased by 15% in the last two decades. Today, one out of every three people is faced with drought and water shortage risk, thus water scarcity and water stress. This book presents studies on the various forms and severity of the drought that can occur in almost every region of the world as well as their causes and impacts. It analyzes in detail the complex drought phenomenon, which has a significant impact on water resources, agriculture, energy production, human health, and forest fires.

This edited collection consists of four sections and ten chapters. The first section discusses the general concept of hydrological drought. The second section examines the effects of climate change on drought and life. The third section evaluates plant behaviors under drought conditions. The fourth and final section investigates the effects of climate change on drought and therefore on life.

Murat Eyvaz
Associate Professor,
Department of Environmental Engineering,
Gebze Technical University,
Kocaeli, Turkey

Ahmed Albahnasawi, Mesut Tekbaş and Ercan Gürbulak
Gebze Technical University,
Kocaeli, Turkey

Section 1

Hydrological Drought

Chapter 1

Hydrological Drought Index Based on Discharge

R. Rintis Hadiani, Bambang Suharto, Agus Suharyanto and Suhardjono

Abstract

Drought is a natural phenomenon causing disasters and its period of occurrence can be predicted in recent times based on several methods using the same or different variables. The prediction is usually associated with the climate interactions in the form of rainfall or discharge patterns which can be analyzed using the return period. Therefore, this research was conducted in four different stages of data acquisition and validation, drought analysis method based on the data, drought prediction method based on hydrology, and sample applications to determine the debit availability in other watersheds. Historical rainfall data converted to dependable rainfall at 80% probability were used as input for the rainfall-discharge analysis while the hydrological drought analysis was conducted using the drought threshold value. Moreover, the drought was predicted using an artificial neural network model while historical data were used to verify the hydrological character of the prediction model. The results of the analysis conducted were further used to predict the water balance in different river areas due to the fact that each area has a different hydrological character. Meanwhile, the watersheds used as case research showed that the model has reliability of up to 80%.

Keywords: drought, rainfall pattern, discharge pattern, hydrological drought, drought index

1. Introduction

Drought is a condition when there is very little extreme rainfall or no rainfall for a relatively long period outside the dry season [1, 2]. Its occurrence in a watershed is recommended to be analyzed using discharge data in addition to the rainfall data [3]. Moreover, the hydrological drought index (HDI) is normally used to describe the drought severity based on discharge data which is indicated as HDI_1, and simulation which is denoted as HDI_2. The discharge prediction model, which is known as the Artificial Neural Network model, is based on a flexible mathematical structure and has the ability to identify a complex nonlinear relationship between input and output [4]. The characters of the rainfall and discharge data were also observed to be adaptable to the method. The purpose of this research is to answer the following questions:

1. What is the estimated discharge calculation model based on rainfall data?

2. How can the dry duration and deficit be determined?

3. What is the hydrological drought index (HDI) in the watershed of the research location?

4. What are the HDI criteria in the watershed of the research location?

2. Literature review

2.1 Data statistical analysis

The rainfall data consistency was tested using multiple mass curves [5, 6], while the discharge data normality test was conducted using probability density function (pdf) analysis with the criterion being that normally distributed data will form a linear or straight line. Moreover, the discharge probability was analyzed using the cumulative distribution function (CDF).

The statistical analysis is usually conducted to determine:

1. Threshold (X0) is the limit determined based on the analysis requirements [7] and according to the selected distribution.

2. X0 as Q50 or Q80 such that Q50 is a normal Q with a probability of 0.5 or data median, while Q80 is a dependable Q with a probability of 0.2.

The criteria are divided into two, which are as follows:

- The value is dry (D) and wet (W) when Q80 < Q [7].

- The value is extremely dry (ED) and very dry (VD) when Q < Q80 [8].

2.2 Calculation of hydrological drought index (HDI) in data statistical analysis

The drought index is a comparison of the deficit to the watershed area as indicated in the following Equation [8]:

$$HDI = \frac{deficit}{Areal} \frac{m^3/sec}{km^2} \tag{1}$$

HDI is the hydrological drought index, the deficit is the difference between X_0 and X_t (ten daily), X_0 is the dry threshold, and X_t is the ten daily periods of discharge.

Drought severity involves the analysis of the duration and deficit in dry conditions.

2.3 Hydrological modeling with artificial neural network (ANN) model

The rainfall and runoff data were simulated through the artificial neural network method using several inputs and outputs [9, 10]. This method imitates the function of the human nervous system and works in line with human learning patterns [11]. The backpropagation using a binary sigmoid activation function has been discovered to be a good ANN model for hydrology. This is due to the fact that the activation function is the net (network) of the linear combination of the inputs and their weights. It is important to note that this model fits the pattern characteristics of the input and output data which are required to follow a normal distribution in the range of 0–1 (0,10,9) [12, 13].

The determination of the hydrological data requires that the parameters recorded for January be influenced by the hydrological conditions in January of the previous year and the same trend is expected to continue from February to December. The data were also sorted from January to December at the end of the time series and the estimation was continued for the next few years. Moreover, the runoff model was designed in line with the rainfall data input based on the analysis of variables in the HDI analysis. Meanwhile, the architecture and equations of the backpropagation method used in this research are presented in **Figure 1** [14, 15].

Where P_1 is the 1st data input, P_n is the nth data input, $Z_{1.1}$ is the 1st auxiliary variable in hidden layer 1, $Z_{1.2}$ is the 2nd auxiliary variable in hidden layer 1, $Z_{2.1}$ is the 1st auxiliary variable in hidden layer 2, $Z_{2.2}$ is the 2nd auxiliary variable in hidden layer 2, b (=1) is the specified bias value which is equal to one, and Qn is the nth data output.

$$z\_net_j = v_{j0} + \sum_{i=1}^{n} P_i v_{ji} \tag{2}$$

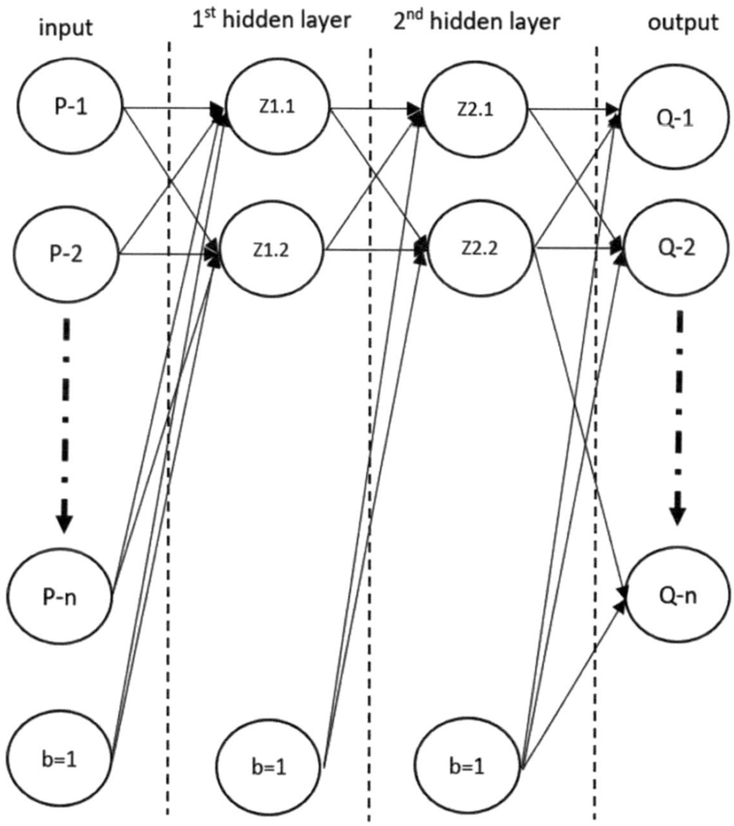

Figure 1.
ANN for discharge calculation.

$$z_j = f\left(z\_net_j\right) = \frac{1}{1 + e^{-z\_net_j}} \tag{3}$$

$$Q\_net_k = w_{ko} + \sum_{j=1}^{p} z_j w_{kj} \tag{4}$$

$$Q_k = f(Q\_net_k) = \frac{1}{1 + e^{-Q\_net_k}} \tag{5}$$

2.4 Discharge prediction model analysis

The results from the model analysis were examined based on the theory of runs which involves the analysis of two processes of sequential and opposite events at a boundary which is within a certain period [16].

The two kinds of "runs" available are run-length and run-sum with the run-length usually used for hydrological analysis related to the prediction of time which is also known as the frequency analysis while the run-sum is for those related to the prediction of rainfall duration and intensity within the analysis season. The overview of these runs is presented in the following **Figure 2**.

The run test was conducted on the simulated data, and since m is a positive run-length while n is a negative run-length as indicated in **Figure 2**, the total run-length (r) is m + n. Moreover, the estimated value of r, E(r), was expressed using the probability (q) for the estimated positive run-length, m, and (p) as the estimated probability value for the negative run-length, n. This relationship was expressed as follows [16]:

$$E\left(r_q\right) = \frac{1}{q(1-p)} = \frac{1}{pq} \tag{6}$$

with boundary conditions $0 < q < 1$

$$p = \frac{m}{m+n}; = \frac{n}{m+n}; \bar{r}_q = \frac{1}{k_r} \sum_{j=1}^{kr} r_{q,j} \tag{7}$$

where $\bar{r}_q$ is the total run-length by q (probability), j = 1, 2, 3, … , kr; and kr is the total number of run-length.

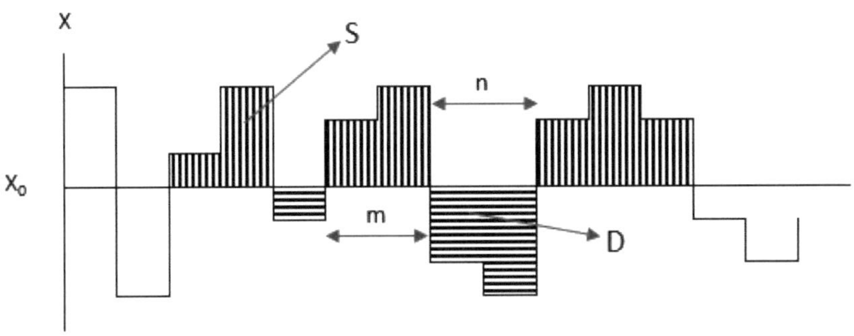

Figure 2.
Overview of positive run-length, m, positive run-sum, S, negative run-length, n, and negative run-sum, D, on a discrete series [16].

It is also important to note that the estimated output in the model analysis is expected to be within the tolerance limit formulated as follows:

$$\frac{1}{pq}-\frac{t_{\alpha/2}}{pq}\left(\frac{p^3+q^3}{k_r}\right)^{1/2}\leq \bar{r}_q \leq \frac{1}{pq}+\frac{t_{\alpha/2}}{pq}\left(\frac{p^3+q^3}{k_r}\right)^{1/2} \tag{8}$$

where α is the tolerance limit (5%) and t is the normal distribution value from the "t-table."

2.5 Drought index parameter reliability analysis

The parameter reliability analysis was conducted because the model describes the sample to be generalized to the population and this led to the application of a sample-based model suitability test. In mathematical analysis, reliability is the ratio of the

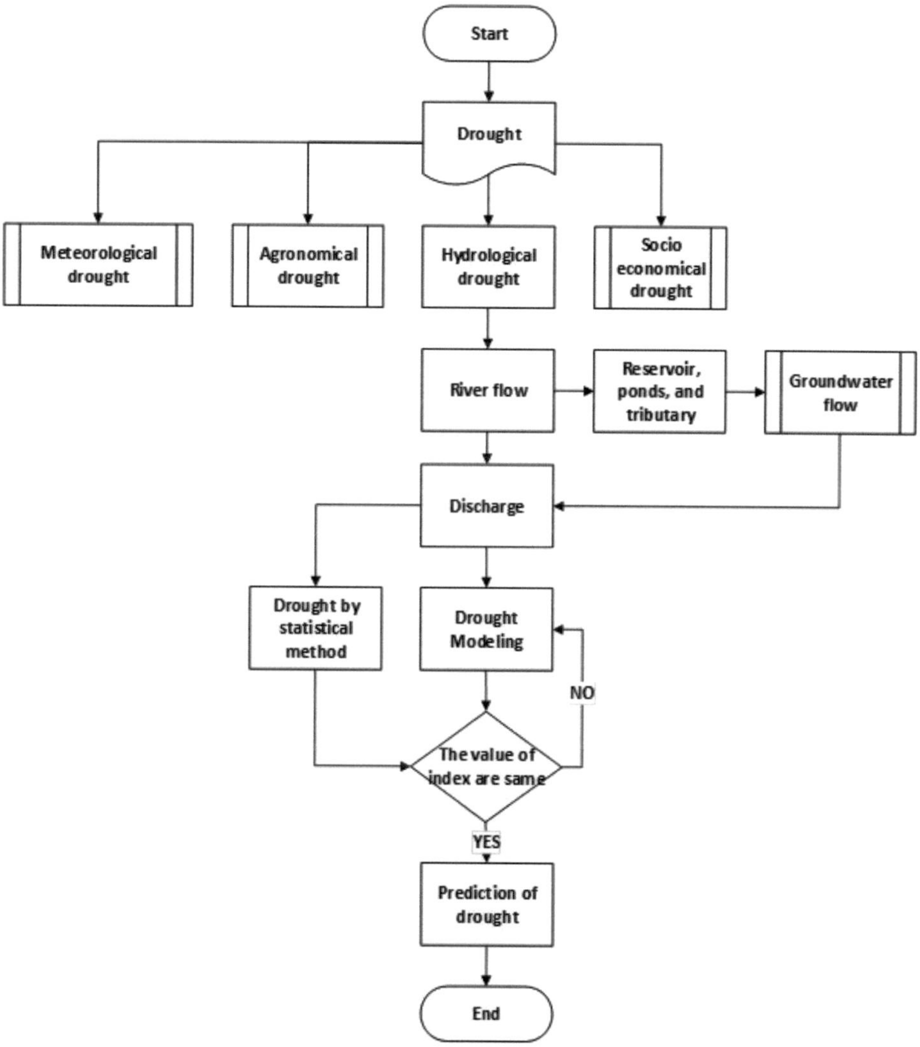

Figure 3.
Concept of research.

total items to the total variance and was applied to the dry duration and deficit variables of the applied watershed. The general formula used is the Cronbach's alpha equation as follows [17]:

$$\alpha_r = \frac{k}{k-1}\left[1 - \frac{\Sigma\sigma_i^2}{\sigma_i^2 + 2(\Sigma\sigma_{ij})}\right] \tag{9}$$

Where $= \frac{1}{10}i^2$ is the number of variances i (the number of diagonals), $\frac{1}{10}ij$ is the covariance of items i and j, $\sigma_i^2 + 2(\Sigma\sigma_{ij})$ is the total variance, and α_r is the reliability of the model.

3. Research method

The rationale of the research concept was used with the drought processes observed to have occurred sequentially starting from meteorological drought, agronomical drought, hydrological, to socioeconomical droughts, as indicated in **Figure 3** [18–20]. The concept was further used to create an operational research

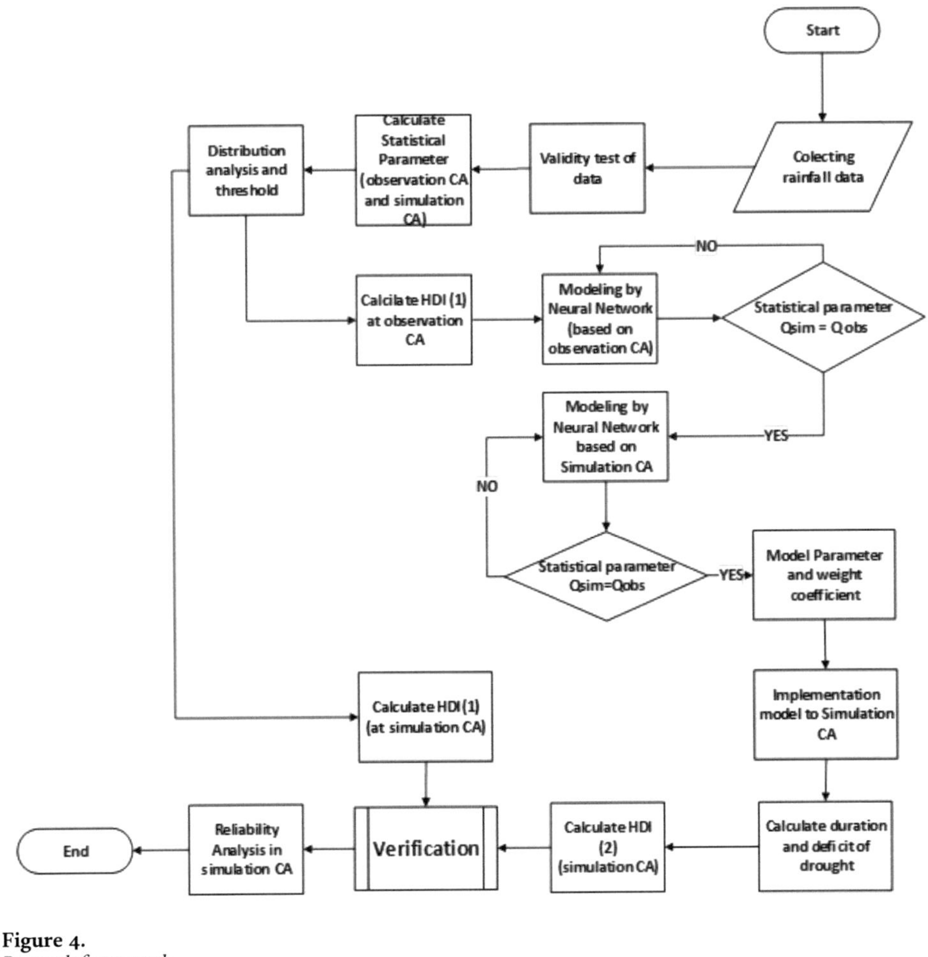

Figure 4.
Research framework.

framework that starts from the preparation of the model through the collection and processing of rainfall, discharge, and climate data of the research location followed by the modeling and verification as well as the calculation of HDI. The model was later applied and tested on the selected watershed as presented in **Figure 4**. The model is built based on historical data in the observation Catchment Area. Then applied to the simulation Catchment Area.

4. Results and discussion

4.1 Data consistency and normality

The data from the rainfall stations in the selected watershed area, including Kedungsangku, Senduro, and Ranupakis stations, were tested for consistency and the results showed that they are consistent. The same process was conducted on the rainfall data from the applied watershed, such as Dampit, Poncokusumo, Sengguruh, Tangkil, and Wagir stations, and the results also showed that the data are consistent.

The normality test was conducted on the discharge data using pdf analysis with the indicator for the normal distribution being the formation of a straight line when the data are plotted visually. The results showed that the discharge data at Automatic Water Level Record (AWLR) and Umbul Dam (observation CA) are not normally distributed and the same trend was also observed for the data from Sutami Dam (Simulation CA).

4.2 Data statistical analysis and threshold

The distribution analysis showed that the discharge data from AWLR and Umbul Dams follow the gamma distribution and the same was also reported for the rainfall data from Kali Asem Catchment Area and Umbul sub-watersheds as well as the discharge data from Sutami Dam.

The next analysis conducted based on the gamma distribution showed the following results for the Kali Asem and Umbul sub-watersheds:

- Qaw50 = 16.27(m^3/sec) which is interpreted as normal discharge in AWLR.
- Qd50 = 21.31(m^3/sec) which is interpreted as normal discharge in Umbul Dam.
- Qaw80 = 11.31(m^3/sec) which is interpreted as dependable discharge in *AWLR*.
- Qd80 = 14.32(m^3/sec) which is interpreted as dependable discharge in Umbul Dam.
- P50 = 36.84(mm) which means normal rainfall.

The same method was used to obtain the threshold values for Sutami Dam as follows:

- P50 = 30.68(mm) which is interpreted as normal rainfall.
- Q50 = 64.34(m^3/sec) which is interpreted as normal discharge in Sutami Dam.
- Q80 = 40.03(m^3/sec) which is interpreted as normal discharge in Sutami Dam.

- WL50 = +267.91(m) which is interpreted as the dependable elevation of the water level in Sutami Dam.
- WL80 = +264.29(m) which is interpreted as the dependable elevation of the water level in Sutami Dam.

4.3 Deficit and dry duration observational data

The deficit is the difference between the volume of water shortage and the threshold, while duration is defined as the total time the deficit occurred. The results of the analysis showed that the duration and deficit do not have the same visual pattern and this means the deficit value does not describe the duration value and vice versa.

The debit deficit on the Very Dry criteria (based on Qaw80), consecutive (run-length) (d), and the total deficit (run-sum) was recorded from January 3 to October 3, 1991, with a total deficit of 142,928 (m^3/sec. Ten daily) and 27 ten daily followed by May 1 to Dec 1, 1997 with 15.88 (m^3/sec. Ten daily) and 22 ten daily, and Jan 1 to Aug 2, 2003 with 46,725 (m^3/sec. Ten daily) and 25 ten daily, respectively, as indicated in **Figures 5** and **6**.

4.4 HDI_1 and HDI_2 in Kali Asem and Umbul sub-watersheds

The calculation showed that the HDI_1 for the Kali Asem sub-watershed at Qaw_{80} was 0.00, the HDI_1 limit at Qaw_{50} was 0.0180, and at 70% Qaw80 was recorded to be −0.0123. Moreover, the result for the HDI_1 in the Umbul sub-watershed at Qd80 was found to be 0.00, the HDI_1 limits at Qd50 were 0.0200 while at 70% Qd80 was −0.0123.

The discharge prediction simulation model implemented the backpropagation ANN using the following parameters:

1. The model consists of one input, two hidden, and two output layers.
2. The network was formed using Descant Gradient Learning (trainingdm) with *logsig* used for activation in the hidden layer and *purelin* in the output layer.

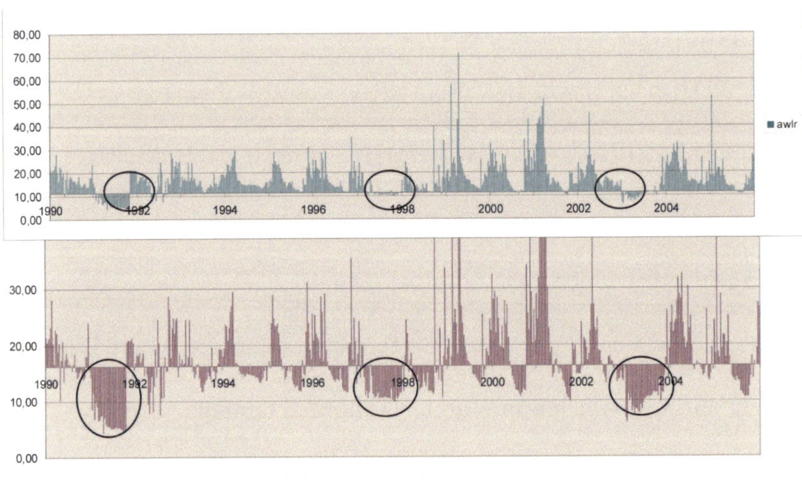

Figure 5.
Deficit based on Qaw50 and Qaw80 in the AWLR of the Kali Asem sub-watershed [21].

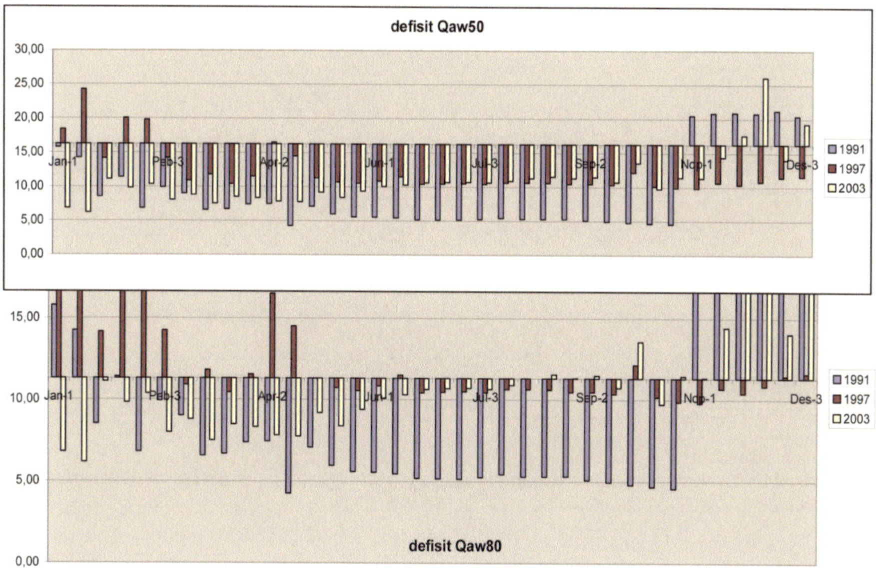

Figure 6.
Deficit of discharge, Qaw in 1991, 1997, 2003 [21].

3. The model simulation stops at the specified epoh of 1000 epoh or a mean square error (MSE) of 0.05.

The "scatterplot" of the $Q_{simulated}$ and $Q_{observed}$ in **Figure 7** shows a straight line and coincides. This means the simulation data has a statistical character that is not significantly different from the observation data.

The HDI_1 value also ranged from 0.002 to 0.024 while the HDI_2 value ranged from 0.001 to 0.022 (Extremely Dry). Moreover, a shift was observed in the dry time and

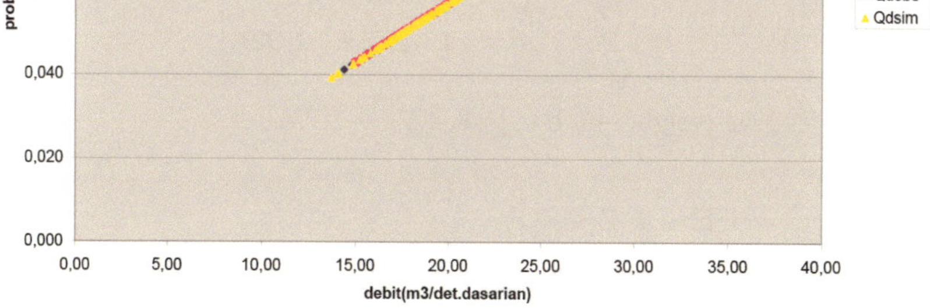

Figure 7.
Suitability analysis for the simulation results based on the discharge at Umbul dam [21].

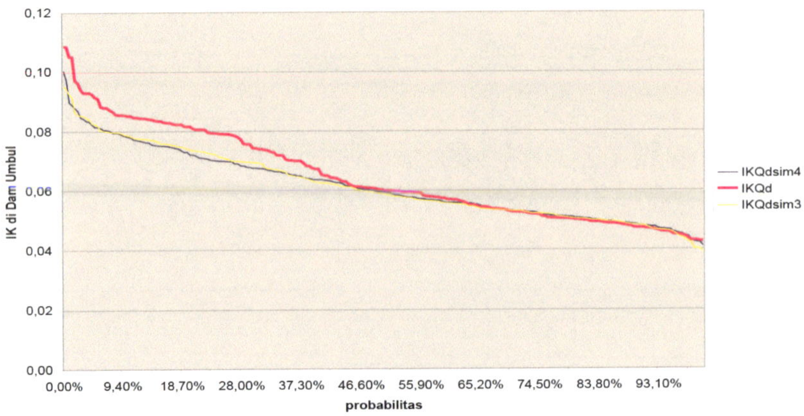

Figure 8.
FDC analysis on the Q observation and simulation ($Q_{observation}$ and $Q_{simulation}$, as well as IK observation and IK simulation [21].

this means it is difficult for the simulation to accurately predict the start or end time for the drought condition (see **Figure 8**).

The calculation showed that the equation can be written in the form of a matrix as follows:

- Input weight, $(P \rightarrow Z_1) = \begin{bmatrix} -2,434 & 2,385 \\ -1,422 & -1,312 \\ -2,730 & -1,428 \\ 2,198 & -2,700 \\ 1,798 & -0,613 \\ -0,823 & 3,194 \\ 1,206 & -0,166 \\ -4,729 & -3,937 \end{bmatrix}$

- Input bias weight, $b = [5,893 \;\; 5,369]$

- Hidden layer weight $- 1, (Z_1 \rightarrow Z_2) = \begin{bmatrix} 0,755 & 7,685 \\ -7,359 & 2,927 \end{bmatrix}$

- Hidden layer bias weight $- 1, b = [-8,477 \;\; -1,755]$

- Hidden layer weight $- 2 \, (Z_2 \rightarrow Q) =$

$$\begin{bmatrix} 0,021 & -0,033 & 0,891 & 0,762 & -0,087 & 0,068 & 0,748 & -0,101 \\ -0,022 & 0,562 & -0,739 & 0,987 & 0,999 & -0,648 & -0,179 & 0,332 \end{bmatrix}$$

- Hidden layer bias weight – 2 b =
$[0,713 \quad 0,129 \quad 0,236 \quad 0,267 \quad 0,681 \quad 0,644 \quad 0,389 \quad 0,383]$

The model was simulated at Umbul Dam for verification and the results showed that the simulation data is within the tolerance limit (α) of 5% which indicates that it is not significantly different. A run test was also conducted on the drought parameter.

- The statistical tests conducted using analysis of variance showed that the statistical value in Levene's test for the data that was not necessarily normally distributed was 1.38 with $p = 0.123$. The p-value $\alpha > (0.05)$ means the null hypothesis was not rejected and this indicates the same variance.

- The Bartlett test also showed that $p = 0.995 > \alpha = 0.05$ and this means the simulation data are not the same as the observation data while the median analysis conducted using Levene's test also obtained $p = 0.123 > \alpha$ (0.05) and this means the null hypothesis was not rejected (same median). Moreover, the run analysis showed that E(r) Qdsim4 was 4.049 and the E(r)Qdsim was 4.016, which lies within the tolerance limit of 4.701–3.331 at α of 5%. This means the simulation data are not the same as the observation data.

4.5 HDI_1 in Sutami watershed

The threshold analysis for the Sutami sub-watershed using observational data from 1991 to 2006 showed the value for Q50 was 64.34 m^3/sec and Q80 was 40.03 m^3/sec. Meanwhile, HDI_1 for Q80 was calculated to be 0.00 while the limit at Q50 was 0.0119 and at 70% Q80 was −0.0059.

The calculation showed that HDI_1 ranged from 0.000 to 0.007 while HDI_2 was from 0.000 to 0.007 in line with the severe drought criteria and this means the simulation was unable to determine the exact time the drought starts and ends. Moreover, the dry deficit duration calculation showed that the simulation data was accepted at $\alpha = 5\%$ while the median test indicated that the simulation data have the same characteristics and are not significantly different from the observation data. The "scatterplot" test of the simulation data presented in **Figure 9** and the Flow Duration Curve test in **Figure 10** also showed that there was a match between the observation and simulation data.

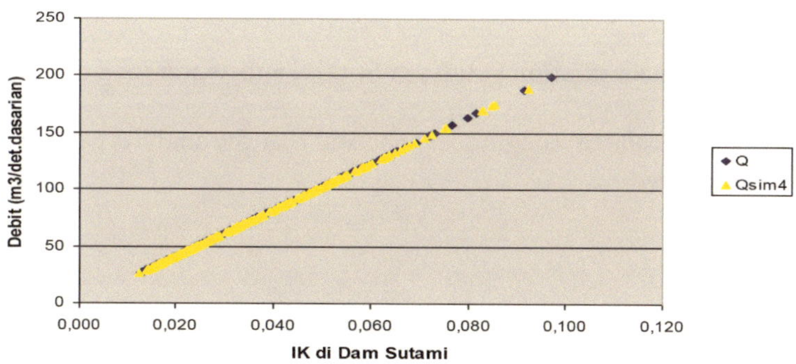

Figure 9.
The suitability analysis of the simulated scatterplot in Sutami dam [21].

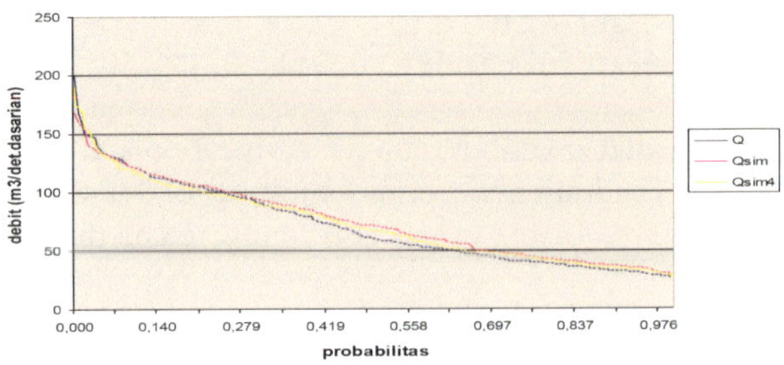

Figure 10.
Flow duration curve conformity analysis of the simulation results at Sutami dam [21].

Variable	N	Mean	StDev	SE-Mean	95% CI		Z	P
Q	288	71,676	35,894	2,115	(67,531;	75,821)	0	0,999
Qsim	288	75,603	33,244	2,115	(71,458;	79,748)	1,86	0,064

Table 1.
Results of the Z-test $Q_{Observation}$ and $Q_{Simulation}$ in Sutami dam [21].

4.6 HDI_2 at Sutami sub-watershed

The Z test analysis showed that $p = 0.064 > \alpha = 0.05$ and this means the simulation and observation are not the same as the mean test while Mann–Whitney median test also showed the value of $p = 0.0547 > \alpha = 0.05$ and this also indicates the same trend as indicated in **Table 1**.

The $Q_{simulation}$ mean value was observed to be within the tolerance range of the $Q_{observation}$ while the p-value $> \alpha = 0.05$ indicates the simulation is not the same as the observation data but the analysis conducted at Sutami Dam showed that the simulation data can be accepted.

The results showed that the simulation can be used to predict the drought in the Sutami sub-watershed for 2007–2014 with the dry limit observed to be at $0.000 < HDI < 0.0119$, the very dry limit at $-0.0059 < HDI < 0.000$, and the extremely dry limit at $HDI < -0.0059$. The analysis further indicates that the lowest HDI value that has ever occurred was -0.006 which is included in the extremely dry criteria.

4.7 Reliability of the simulation results in Sutami sub-watershed

The reliability calculated using Eq. (9) showed that the analysis value for Q_{50} was 79% and Q_{80} was 81%.

5. Conclusion

The research concludes that:

1. The calculation model for the estimated discharge is based on the ten daily rainfall data from the selected watershed which includes Kali Asem Catchment

Area and Umbul sub-watersheds. This is in the form of a mathematical equation which is stated as follows

$\mathbf{Q1} = -0{,}021\ \mathbf{Z_{21}} - 0{,}022\ \mathbf{Z_{22}} + 0{,}713$

where:

$\mathbf{Z_{21}} = 0{,}755\ \mathbf{Z_{11}} + 7685\ \mathbf{Z_{12}} - 8477$

$\mathbf{Z_{22}} = -\ 7359\ \mathbf{Z_{11}} + 2927\ \mathbf{Z_{12}} - 1755$

And,

$\mathbf{Z_{11}} = -2434\ \mathbf{P1} - 1422\ \mathbf{P2} - 2730\ \mathbf{P3} + 2198\ \mathbf{P4} + 1798\ \mathbf{P5} - 0{,}823\ \mathbf{P6} + 1206\ \mathbf{P7} - 4729\ \mathbf{P8} + 5893$

$\mathbf{Z_{12}} = 2385\ \mathbf{P1} - 1312\ \mathbf{P2} - 1428\ \mathbf{P3} - 2700\ \mathbf{P4} - 0{,}613\ \mathbf{P5} + 3194\ \mathbf{P6} - 0{,}166\ \mathbf{P7} - 3937\ \mathbf{P8} + 5369$

$\mathbf{Q2} = -0{,}033\ \mathbf{Z_{21}} + 0{,}562\ \mathbf{Z_{22}} + 0{,}129$

$\mathbf{Q3} = 0{,}891\ \mathbf{Z_{21}} - 0{,}739\ \mathbf{Z_{22}} + 0{,}236$

$\mathbf{Q4} = 0{,}762\ \mathbf{Z_{21}} + 0{,}987\ \mathbf{Z_{22}} + 0{,}267$

$\mathbf{Q5} = -0{,}087\ \mathbf{Z_{21}} + 0{,}999\ \mathbf{Z_{22}} + 0{,}681$

$\mathbf{Q6} = 0{,}068\ \mathbf{Z_{21}} - 0{,}648\ \mathbf{Z_{22}} + 0{,}644$

$\mathbf{Q7} = 0{,}748\ \mathbf{Z_{21}} - 0{,}179\ \mathbf{Z_{22}} + 0{,}389$

$\mathbf{Q8} = -0{,}101\ \mathbf{Z_{21}} + 0{,}332\ \mathbf{Z_{22}} + 0{,}383$

The simulation test showed that the model was accepted at $\alpha = 5\%$ or 95% confidence level and the reliability of its parameters, when applied to Sutami sub-watershed, was found to be only 80%.:

2. The dry duration and deficit were determined based on the threshold value obtained from historical discharge data using pdf and CDF analysis. The criteria used were that the watershed is believed to be dry to wet when the discharge is above the threshold and dry when it is below the threshold. Moreover, the analysis showed that the data were gamma distributed with a threshold value of:

    - P50 = 36.84 which means the normal rainfall in the analysis period was 36.84 mm/ten daily.

    - Qaw50 = 16.27 which means the normal discharge at AWLR during the analysis period was 16.27 m^3/sec. Ten daily.

    - Qd50 = 21.31 which means the normal discharge at AWLR in the analysis period was 16.27 m^3/sec. Ten daily.

    - Qaw80 = 11.31 which means that the dependable discharge in AWLR in the analysis period is 11.31 m^3/sec. Ten daily.

    - Qd80 = 14.32 which means the dependable discharge at Umbul Dam during the analysis period was 14.32 m^3/sec. Ten daily.

3. HDI for Kali Asem and Umbul sub-watersheds have the following limitations:

- Dry when 0.000 < HDI < 0.018,
- Very Dry when −0.012 < HDI < 0.00,
- Extremely Dry when HDI < −0.012.

HDI in Sutami sub-watershed has the following limits:

- Dry when 0,000 < HDI < 0,012,
- Very Dry when −0,006 < HDI < 0,00,
- Extremely Dry when HDI < −0,006.

4. The criteria for IK at Kali Asem and Umbul sub-watersheds based on the research results showed the extremely dry (ED) with duration sharpness in zone 2 which occurred in 1991 and very dry (VD) with a sharpness of zone 2 in 2003. Meanwhile, the lowest HDI value in 1991 was −0.024 and the duration was 23 ten daily. In 2003, the lowest HDI value was −0.016 with a mean value of −0.007 and a duration of 23 ten daily. This means the drought occurred twice during the analysis period, in 1991 and 2003, with a duration sharpness of more than 7 months which exceeds the usual 6 months for the dry season.

Author details

R. Rintis Hadiani[1*], Bambang Suharto[2], Agus Suharyanto[2] and Suhardjono[2]

1 University of Sebelas Maret, Surakarta, Indonesia

2 University of Brawijaya, Malang, Indonesia

*Address all correspondence to: rintis@ft.uns.ac.id

References

[1] Moreland JA. Drought: U.S. Geological Survey Water Fact Sheet. Open-File Report 93–642, 2; 1993 www.md.water.usgs.gov/faq/drought.html. October 10, 2005

[2] van Noordwijk M, Agus F, Suprayogo D, Hairiah K, Pasya G, Verbist B, et al. Peranan Agroforestri dalam mempertahankan Fungsi Hidrologi Daerah Aliran Sungai (DAS). Ağrı. 2004;**26**(1)

[3] Samiee M. A regional model of the hydrological drought Sverity for Tehran Province, European geophysical society 2003. Geophysical Research Abstracts. 2003;**5**(01654):2003

[4] Nugroho AS. The applications to character recognition, meteorological prediction, and bioinformatics problem. In: A Dissertation of Doctor Engineering, Nagoya Institute of Technology. Japan: Information Analysis using Softcomputing; 2003

[5] Sumarto CD. Hidrologi Teknik. Surabaya: Usaha Nasional; 1987

[6] Harto S. Analisis Hidrologi. Jakarta: PT Gramedia Pustaka Utama; 1993

[7] Fleig AK, Tallaksen LM, Hisdal H, Demuth S. A global evaluation of streamflow drought characteristics. Hydrology and Earth System Sciences. 2006;**10**:535-552. DOI: 10.5194/hess-10-535-2006

[8] Tallaksen LM, Van Lanen H. Understanding Droughts. International Water Management Institute; 2005 www.iwmi.cgiar.org/drw/info/default.asp?PGID=2.htm

[9] Wu JS, Han J, Annambhotla S, Bryant S. Artificial neural networks for forecasting watershed runoff and stream flows, abstract ACSE research. Journal of Hydrologic Engineering. 2004;**10**(3): 216-222 (May/June 2005)

[10] Tallaksen LM, Madse H, Clausen B. On the definition and modelling of streamflow drought duration and deficit volume. Hydrological Sciences-Journal-des Sciences Hydrologiques. 1997;**42**(1) www.cig.ensmp.fr/~iahs/hsj_420/hysj_42_01_0015

[11] Reda Taha. Artificial Neural Network (ANNs) in Structural Engineering. 2002. www.reda_taha.com/research/ANN.htm

[12] Siang JJ. Jaringan Syaraf Tiruan dan Pemrogramannya Menggunakan Matlab. Penerbit Andi Yogyakarta; 2005

[13] DARPA Neural Network Study. Neural Network FAQ, Part 1 of 7: Introduction. AFCEA International Press; 1988 www.ftp.sas.com/pub/neural/FAQ.html#A2

[14] Suharyanto. Aplikasi Artificial Neural Network di Bidang Keairan. In: Disampaikan pada Seminar dan Pameran Teknologi Komputer, 22–23 September 1997. Diselenggarakan oleh KSAKS, Senat Mahasiswa Fakultas Teknik, UNDIP; 1997

[15] Kim TW, Valdes JB. A nonlinear model for drought forecasting based on conjunction of wavelet transforms and neural networks. Journal of Hydrologic Engineering, ASCE. 2003 www.u.arizona.edu/~jvaldes/Fcst/Asce/Final.pdf

[16] Yevjevich V. Stochastic Processes in Hydrology. Fort Collins, Colorado, USA: Water Resources Publications; 1972

[17] Zulganef. Pemodelan Persamaan Struktur & Aplikasinya Menggunakan

Amos 5. Bandung: Penerbit Pustaka; 2006

[18] Jacobsen HL. Drought Outlook. U.S. Drought Outlook: National Drought Mitigation Center; 2007. www.drough toutlook.com

[19] Beste D. Drought. Watershed TV; 2004. www.greentreks.org/watershedtv/ intheflow/drought.htm. January 7, 2005

[20] Tallaksen, LM. A Global Evaluation of Streamflow drought Characteristics. 2006. http://www.hydrol-earth-syst-sci. net/10/535/2006/hess-10-535-2006.pdf

[21] R. Rintis Hadiani. Drought Analysis based on Hydrological Data. 2009. Dissertation. https://prasetya.ub.ac.id/ en/disertasi-rintis-hadiani-analisis-kekeringan-berdasarkan-data-hidrologi/

Chapter 2

Review of Hydrological Drought Analysis Status in Ethiopia

Kassa Abera and Admasu Gebeyehu

Abstract

Drought is a complex natural disaster unlike flood, which covers a large area when it occurred. This review was conducted on hydrological drought analysis and monitoring status in Ethiopia by reviewing the master plan of eight major river basins and previous research related to drought. A total of 24 article papers was reviewed and it is found that hydrological drought analysis studies cover only 8.33% of all of the river basins in Ethiopia. Researchers in the region have focused primarily on meteorological drought (37.5%) rather than hydrological and agricultural drought analysis. Although Ethiopia has long been dependent on rainfed agriculture for its economy and remains the primary livelihood of the population, the Ethiopian government has begun focusing on transitioning to an industrial economy, placing pressure on the water resource. In a region plagued by drought, drought analysis, and monitoring, drought early warning systems and effective mitigation measures are still limited and even lacking in some areas. Therefore, emphasis on hydrological drought analysis and development of suitable drought mitigation measurements is important to implement strategies for effective and sustainable water resource management by which water may remain available during the long dry seasons and the impacts of hydrological drought may be lessened.

Keywords: hydrological drought, drought mitigation, Ethiopian river basins

1. Introduction

Drought is a worldwide natural hazard and has a detrimental impact on society, the environment, and the economy [1]. Extreme hydrological events both high (flood) and low (drought) flow are of particular concern globally. Of these hydrological extremes, drought is the most complex and widespread [2]. It is one of the most common natural events that has devastating negative impacts on agriculture and water resources [3].

There is no universal definition for drought due to its complexity [4]. Therefore, meteorologists defined drought as a scarcity of precipitation [5–10]; hydrologists have defined hydrological drought as scarcity of surface and subsurface water [5, 6, 7–15]; agriculturalists and agronomists defined agricultural drought as related to soil moisture deficiency [3, 16, 17] and sociologists and economists defined the overall welfare crisis of the society caused by drought to be socioeconomical drought [4, 18–21]. These types of droughts have accumulating effects, thus meteorological drought results in losses, such as crop stress, predation by pests, and disease due to low moisture, to the agricultural systems while hydrological drought causes the shortage of water supply, decrease in reservoir water level and groundwater volume, lower irrigation and hydropower production [14]. The accumulation of meteorological and hydrological drought results in socioeconomical drought in which the overall ecosystem will be disturbed and human and animal lives will be negatively impacted and even lost [15].

Historically, Ethiopia has faced multiple seasonal drought events due to erratic rainfall and climate change [22]. The most drought-prone areas in Ethiopia are in Northeast Ethiopia and the Upper Blue Nile basin, including the Northern Tigray region, some parts of Amhara regions, such as South Wollo, North Wollo, South Gondar, and Afar Region, most parts of Somalia Region, and Eastern parts of Oromia Region [1, 3, 23–27]. Drought in Ethiopia occurs at a recurrence interval of 3–10 years [1], and even though this frequent recurrence is common, there still lacks any firmly established drought mitigation measure for these events. Only short-term response efforts are provided in the form of food aid when food supplies have decreased significantly due to extended drought.

Meteorological drought analysis has been studied frequently, yet hydrological and agricultural drought analysis and monitoring are not studied adequately. It is thought that Ethiopia is a water tower in East Africa but water resource management over the region is not well developed. This aggravates the natural hazard, such as drought impact on human life. Hydrological drought has a great influence on water supply irrigation and power production by reducing the availability of surface and subsurface water. There are few dams and reservoirs in the country and most of them are hydropower plants. But there is a lack of water conservation to reduce drought impact when it occurs. Generally, drought monitoring and forecasting studies are untouched and need a thorough investigation to alleviate socioeconomic problems related to drought.

The objective of this review chapter is to assess the status of hydrological drought studies in Ethiopia by reviewing different previously studied article papers related to drought. A total of 24 article papers was reviewed and the master plan of the eight-river basin was also reviewed. Of these, only two papers were related to hydrological drought and the remains were about meteorological and other drought-related topics. This implies that hydrological drought studies in Ethiopia require further analysis, monitoring, and forecasting investigation. Therefore, it is important to do this kind of review to show the gap of drought studies over the region for future researchers, stakeholders, and planners to develop a suitable early warning system.

2. Materials and methods

2.1 Description of the study area

Ethiopia has an ample amount of water resources when compared to other African countries yet the development is still poor. There are 12 major river basins in the country which generate an annual runoff of 123 BM^3 (**Table 1**). From these, Aysha and Ogaden river basins are dry and the Mereb and Denakle have insignificant streamflow over the year, the border basins from North to East direction (**Figure 1**). Eight river basins have a well-organized master plan, however, only the three river basins (Abbay, Awash, and Tekeze) are popularly studied for the development of

River basin	Area (km^2)	Annual runoff (BM^3)	Terminus
Abbay	199,912	52.6	Mediterranean
Awash	110,000	4.6	Within the country
Baro	75,912	23.6	Mediterranean
Genale Dawa	172,259	5.8	Indian Ocean
Omo Gibe	79,000	17.9	Lake Turkana
Tekeze	82,350	7.6	Mediterranean
Rift Valley	52,000	5.6	Chew Bahir
Wabishebele	202,220	4.6	Indian Ocean
Mereb	5900	0.26	Sudanese Wetland
Denakle	64,380	0.86	Within the country
Aysha	2223	0	
Ogaden	77,120	0	
Total	**1,123,276**	**123.42**	

Source: River Basin Master Plan; Ministry of Water, Irrigation and Electricity, Ethiopia.

Table 1.
Characteristics of Ethiopian major river basins.

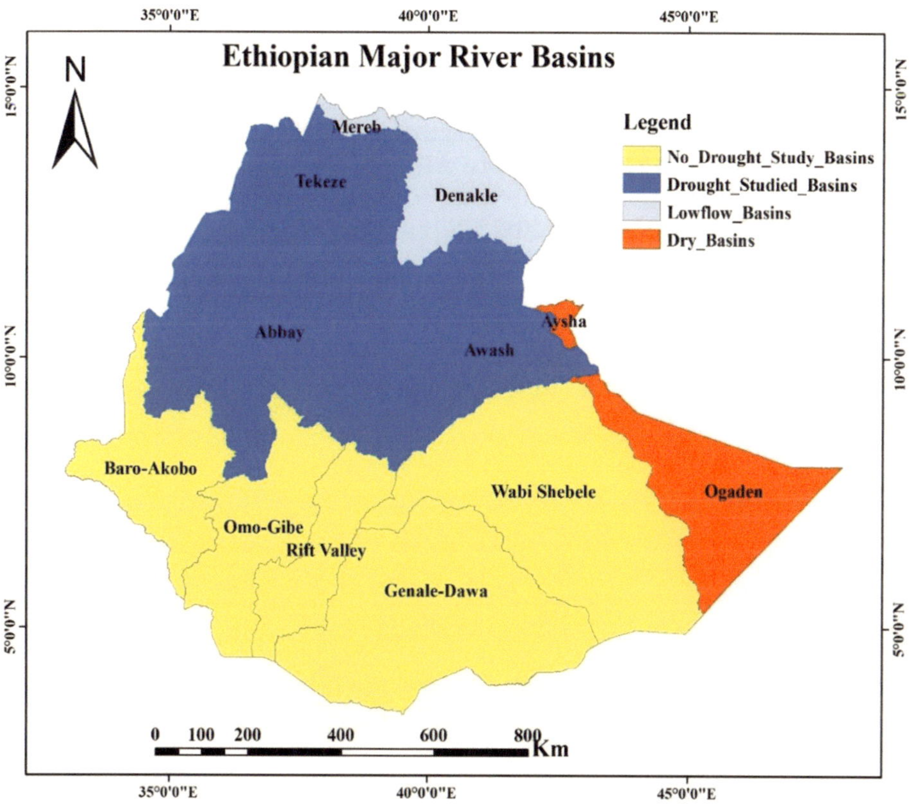

Figure 1.
Drought study information of Ethiopian river basins.

irrigation, water supply, and hydropower projects. Different types of drought studies were also relatively studied in these river basins. In the Wabishebele river basin, one hydrological drought analysis was studied by Awas [26]. Abbay and Awash basins have good hydrometeorological data and are highly invested when compared to other river basins. This review is focused on the assessment of hydrological drought analysis and the drought mitigation approach of previous research in Ethiopia, related to drought.

Spatially, the Abbay river basin is the largest and it covers 43.1% of the surface runoff of the country. The general characteristics of each river basin in the country are given in **Table 1**. In Ethiopia, there is a high seasonal flow and rainfall variation. As shown in **Figures 2** and **3**, Abbay and Omo gibe river basins have a high flow when compare to other river basins and overall the maximum flow is obtained during the summer season from June to August (JJA).

Ethiopia has 12 major river basins, most of which are transboundary rivers except the Awash river. The total surface water is estimated at 124 BM^3 and the groundwater potential is estimated near 30 BM^3 [28]. Up to 70% of the surface water is originated from the central and western highlands on the western sides of the Great Rift Valley flow to the west into the Nile river basin system that covers 39% of the landmass and the remaining 30% of surface water originated from eastern highlands flow into east that covers 61% of the landmass.

2.2 Historical drought in Ethiopia

Ethiopia is experienced severe drought problems for the last decades. According to Mohammed et al., the most drought years in North East Highlands of Ethiopia were 1984, 1987, 1988, 1992, 1993, 1999, 2003, 2004, 2007, and 2008 [1]. Bayissa et al. also found that 1984/85 and 2003/04 were the extreme drought years in the Upper Blue Nile basin in Ethiopia [29]. Based on EM-DAT, 2014, the most severe drought years in Ethiopia from 1900 to 2013 were 1965, 1969, 1973, 1983, 1987, 1989, 1997, 1998, 1999, 2003, 2005, 2008, 2009, and 2012 with an average recurrence interval of 4 years

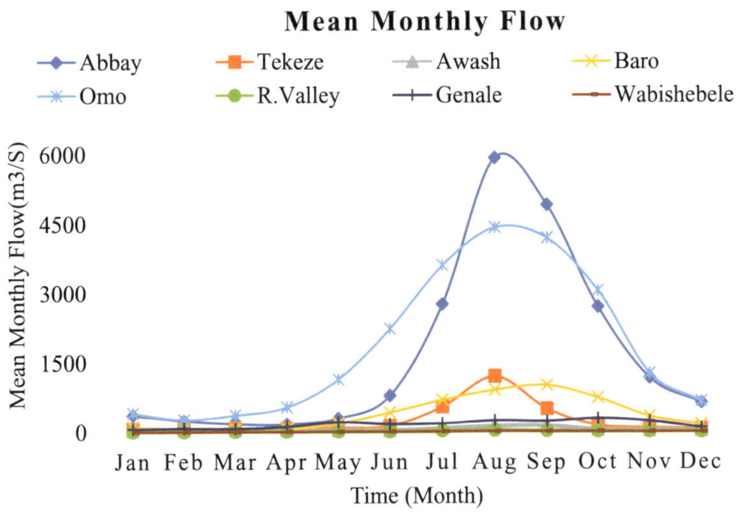

Figure 2.
Seasonal variation of streamflow over Ethiopian river basins.

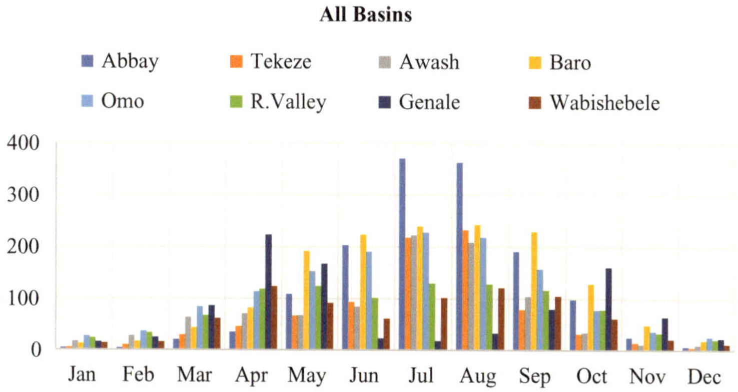

Figure 3.
Mean monthly rainfall of eight river basins in Ethiopia.

No.	Author	Title	Drought Category
1	Philip et al. [22]	Attribution analysis of the Ethiopian drought of 2015	General
2	Belayneh et al. [2]	Long-term SPI drought forecasting in the Awash river basin in Ethiopia using wavelet neural network and wavelet support vector regression models	Meteorological
3	Yimer et al. 2017	Meteorological drought assessment in northeast highlands of Ethiopia	Meteorological
4	Araya and Leo Stroosnijder, 2011	Assessing drought risk and irrigation need in northern Ethiopia	General
5	Enyew et al. [27]	Assessment of the impact of climate change on hydrological drought in Lake Tana catchment, Blue Nile basin, Ethiopia	Hydrological
6	Edosa et al., 2010	Drought analysis in the Awash river basin, Ethiopia	Hydrometeorological
7	USAID Report, 2018	Economics of resilience to drought; Ethiopia analysis	Socioeconomic
8	Philip et al. [22]	The drought in Ethiopia, 2015	General
9	Jjemba et al.	Extreme drought in Ethiopia stretches drought management systems	Socioeconomic
10	Gebrehiwot et al. [24]	Spatial and temporal assessment of drought in the Northern highlands of Ethiopia	Meteorological
11	Bayissa et al. [17]	Comparison of the performance of six drought indices in characterizing historical drought for the Upper Blue Nile basin, Ethiopia	Meteorological
12	Awass [26]	Hydrological drought analysis occurrence, severity, risks: the case of Wabishebele river basin, Ethiopia	Hydrological
13	EL Kenawy et al., 2016	Changes in the frequency and severity of meteorological drought over Ethiopia from 1960 to 2013	Meteorological

No.	Author	Title	Drought Category
14	Bayissa et al. [29]	Spatio-temporal assessment of meteorological drought under the influence of varying record length: the case of Upper Blue Nile basin, Ethiopia	Meteorological
15	Zeleke et al. [18]	Trend and periodicity of drought over Ethiopia	Meteorological
16	Teshome and Zhang [20]	Increase of extreme drought over Ethiopia under climate warming	General
17	Viste et al. [19]	Recent drought and precipitation tendencies in Ethiopia	General
18	Getachew et al., 2020	Application of artificial neural networks in forecasting a standardized precipitation evapotranspiration index for the Upper Blue Nile basin	Meteorological
19	Getachew, 2018	Drought and its impacts in Ethiopia	Socioeconomic
20	Temam et al., 2019	Long-term drought trends in ethiopia with implications for dryland agriculture	Agricultural
21	Dawit et al., 2019	Comparison of meteorological and agriculture-related drought indicators across Ethiopia	Meteorological and agricultural
22	Y.A. Bayissa et al., 2018	Developing a satellite-based combined drought indicator to monitor agricultural drought: a case study for Ethiopia	Agricultural
23	IDA GRANT-H0280, 2011	Emergency drought recovery project (EDRP) in Ethiopia	General
24	Sara Pantuliano and Mike Wekesa, 2008	Improving drought response in pastoral areas of Ethiopia	General

Table 2.
Summary of selected literature related to drought studies in Ethiopia for this review.

[30]. Generally, the year 1984 was a bad drought event in Ethiopia and it was globally known. Here, all the above-stated drought years were analyzed based on meteorological drought indicators, especially standardized precipitation index (SPI) and palm drought severity index (PDSI).

2.3 Data collection and analysis

To review the status of hydrological drought conditions in Ethiopia, important data were collected from the Ministry of Water, Irrigation, and Electricity, department of Basin Development Authority. The river basin master plan was thoroughly reviewed and previous drought-related studies in Ethiopia were also assessed.

During this review, 24 articles and conference papers related to drought studies in Ethiopia were collected. From these, nine papers are meteorological drought studies, seven papers are general drought impact studies, and the remaining eight were agricultural, hydrological, and socioeconomic drought studies (**Tables 2** and **3**). Surprisingly, except for some general drought studies related to drought impact over the country, other drought studies were conducted in some specific parts of the

NO.	Basin	Article related to meteorological drought	Articles related to hydrological drought
1	Abbay	3	1
2	Awash	2	
3	Omo-Gibe	1	
4	Rift Valley	1	
5	Tekeze	2	
6	Wabishebele		1

Table 3.
Different types of drought studies status in each river basin.

country. Especially meteorological drought studies were highly focused on the Abbay river basin (Upper Blue Nile) and Awash river basin. Agricultural and socioeconomic drought studies slightly tried to see the overall drought conditions in Ethiopia. However, these are also not studied in-depth.

Agricultural and socioeconomic drought studies were not focused on a particular river basin. Total 13 articles, including agricultural, socioeconomic, and general concepts, and drought impacts in Ethiopia were covered in some parts of the country without specifying a particular river basin.

3. Result and discussion

3.1 Hydrological drought status of the country

Ethiopia has been affected by drought many times over the last few centuries. However, drought studies and mitigation measurement investigation are still limited. Although there are few drought studies in the country; it is insufficient. Especially agricultural, hydrological and socioeconomic drought studies are untouched. As shown in **Table 4** and **Figure 4**, most drought studies in Ethiopia are focused on meteorological drought and other general drought-related impact assessments. Meteorological drought is highly varying within the short-period scale in a month depending on the precipitation variability. Therefore, drought analysis from a short-time scale may lead to an erroneous conclusion. But hydrological drought study

Type of drought	Number of studies	Percentage (%)
Meteorological drought	9	37.5
Hydrological drought	2	8.33
Agricultural drought	3	12.5
Socioeconomic drought	3	12.5
General related to drought impact	7	29.16
Total articles reviewed	24	100

Table 4.
Types of drought studies over Ethiopia.

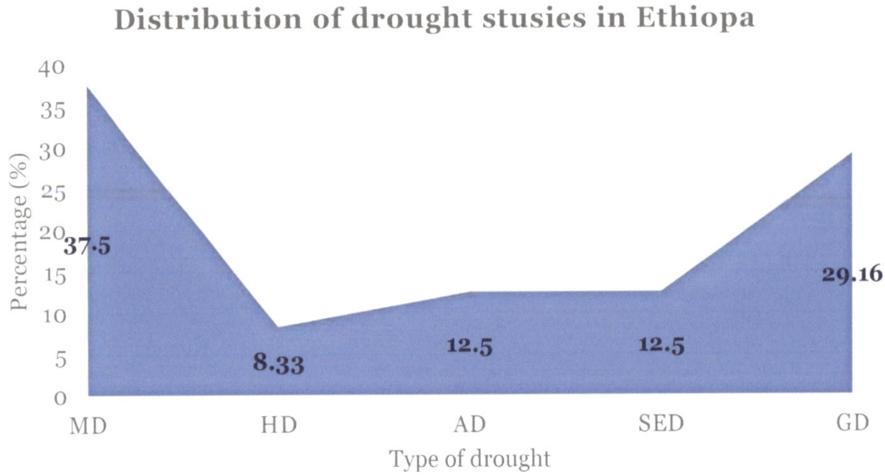

Figure 4.
Percentage of drought studies in Ethiopia (MD = meteorological drought, HD = hydrological drought, AD = agricultural drought, SED = socioeconomical drought, and GD = general drought-related studies).

requires a long-term time scale greater than 6-month cumulative drought conditions of the study area. Mostly hydrological drought analysis is conducted annually based on and above, which will give some concrete information about the drought situation of a particular study area. From this review, hydrological drought studies were covered only 8.33%, which implies that it needs further study (one article in Abbay subbasin and one article from Wabishebele basin). Almost 78% of the study were concentrated in North Eastern and Upper Blue Nile basin, Tekeze and Abbay, and Awash river basin and which is meteorological drought (**Table 3**). Two researchers have been studied, hydrological drought in Abbay and Wabishebele basins (**Table 3**). But the remaining six basins are still not studied. Now the government of Ethiopia is planning to transform from agricultural lead to industrial transformation. This will have achieved when the natural resource will be properly managed and utilized. Water is the central part of all infrastructures development. However, the master plan of major river basins in Ethiopia focused only on the potential assessment of irrigation and hydropower, and there is no drought trend analysis and future hydrological drought forecasting. Hydrological drought affects irrigation, water supply, hydropower, and other water-related sectors. So, it is important to study the historical hydrological drought characteristics, such as frequency, magnitude, duration, severity, and future probability of the basin streamflow to satisfy all demands.

As far as reviewed from the basins master plan report and previous pieces of literature, there is no method adopted to analyze the hydrological drought in the region. But for sustainable water resource development, mitigation measurements of the extreme hydrological events, such as floods and drought, are impropriated. Otherwise, simply constructing any structure in the basin alone may not be a solution to improve poverty over the country.

3.2 Meteorological and agricultural drought

From the reviewed papers, 37.5% was covered meteorological drought analysis and monitoring studies, and agricultural drought studies were covered 12.5% (**Table 4**). Ethiopia is highly dependent on rainfed agriculture; so, meteorological and agricultural drought analysis, monitoring, and early warning system development are crucial. But still, there is no well-adopted drought analysis technique for a nationwide or a regional level. As a result, the development of drought early warning system has lacked. At the same time, hydrological drought analysis and monitoring is also key point for river basin development and water resource management. But due to its large input data requirement, hydrological drought study is not further investigated.

3.3 General drought-related studies

The socioeconomic of Ethiopia is continuously affected by frequent drought disasters. It is difficult to cope with subsequent years after drought has occurred. Up to 29.16% of the reviewed papers were related to drought impact, attribution, economics resilience to drought, extreme drought assessment, trend, and periodicity of drought in Ethiopia [4, 18–20]. Except for some articles, most of the reviewed articles were conducted in some parts of the country and did not give good information about the effect of drought in the country.

4. Conclusion

During any river basin master planning, considering extreme hydrological events, such as floods and drought, are the important issues for sustainable water resource development. Otherwise, simply focusing on the investigation and assessment of the available natural resources in a specific river basin and utilization of the resource will never bring development. Particular attention is to be given to drought-affected areas and conjunctive use of ground and surface water is encouraged. Aridity is the general characteristic of an arid climate and represents a (relatively) permanent condition, while drought is temporary. In an arid climate, drought can still occur when local conditions are even drier than normal. But 90% of the reviewed studies in Ethiopia were conducted on arid and semiarid areas of the region. Generally, hydrological drought study lacked in the country. Therefore, in the future, it is important to focus on hydrological drought monitoring and forecasting to achieve the sustainable utilization of available water resources in Ethiopia.

Acknowledgements

All the river basin master plan documents were freely accessed from the Ministry of Water, Irrigation, and Electricity of Ethiopia. Therefore, great gratitude is given to all the staff members of the ministry, especially for Basin Development Authority Department.

Conflict of interest

We declared that we have no known competing financial interests or personal relationships that could have appeared to influence the work reported in this chapter.

Author details

Kassa Abera[1,2]* and Admasu Gebeyehu[2]

1 Wollo University, Kombolcha Institute of Technology, KIoT, Ethiopia

2 Addis Ababa University, Addis Ababa Institute of Technology, AAiT, Ethiopia

*Address all correspondence to: kassaabera21@gmail.com

References

[1] Mohammed Y, Yimer F, Tadesse M, Tesfaye K. Meteorological drought assessment in north east highlands of Ethiopia. International Journal of Climate Change Strategies and Management. 2018;**10**:142-160

[2] Belayneh A, Adamowski J, Khalil B, Ozga-zielinski B. Long-term SPI drought forecasting in the Awash river basin in Ethiopia using wavelet neural network and wavelet support vector regression models. Journal of Hydrology. 2014;**508**:418-429

[3] Araya A, Stroosnijder L. Assessing drought risk and irrigation need in northern Ethiopia. Agricultural and Forest Meteorology. 2011;**151**:425-436

[4] Mera GA. Drought and its impacts in Ethiopia. Weather and Climate Extremes. 2018;**22**:24-35

[5] Keskin ME, Terzi Ö, Taylan ED, Küçükyaman D. Meteorological drought analysis using artificial neural networks. Academic Journals. 2011;**6**:4469-4477

[6] Shen H, Yuan FEI, Ren L, Kong HAO, Tong RUI. Regional drought assessment using a distributed hydrological model coupled with standardized runoff index. Remote Sensing and GIS for Hydrology and Water Resources (IAHS Publ). 2015;**368**:397-402

[7] Svoboda MD, Fuchs BA. Handbook of Drought Indicators and Indices. Geneva: World Meteorological Organization (WMO); 2017. ISBN 9781351967525

[8] Pashiardis S, Michaelides S. Implementation of the standardized precipitation index (SPI) and the reconnaissance drought index (RDI) for regional drought assessment: A case study for. European Water. 2008;**23**:57-65

[9] Mohammed Y. Meteorological drought assessment in north east highlands of Ethiopia. International Journal of Climate Change Strategies and Management. 2018;**10**(1):121-141. DOI: 10.1108/IJCCSM-03-2017-0059

[10] Zargar A, Sadiq R, Naser B, Khan FI. A review of drought indices. NRC Research Press. Environmental Reviews. 2004;**19**:333-349

[11] Khanna M. Hydrological Drought Indices. New Delhi, India: Water Technol. Centre, Indian Agric. Res; 2010

[12] Trambauer P, Werner M, Winsemius HC, Maskey S, Dutra E, Uhlenbrook S. Hydrological drought forecasting and skill assessment for the Limpopo river basin, Southern Africa. Hydrology and Earth System Sciences. 2015;**19**:1695-1711

[13] Abcdef IWY, Adef MB. Hydrological Drought Index Based on Reservoir Capacity—Case Study of Batujai Dam in Lombok Island. Indonesia: West Nusa Tenggara; 2018

[14] Van Loon AF. Hydrological drought explained. Wiley Interdisciplinary Reviews Water. 2015;**2**:359-392

[15] Boudad B, Sahbi H, Manssouri I. Analysis of meteorological and hydrological drought based in SPI and SDI index in the Inaouen Basin (Northern Morocco). Journal of Materials and Environmental Sciences. 2018;**9**:219-227

[16] Tsige DT, Uddameri V, Forghanparast F. Comparison of

meteorological and agriculture-related drought indicators across Ethiopia. Basel, Switzerland: MDPI Water; 2019

[17] Bayissa Y, Maskey S, Tadesse T, van Andel SJ, Moges S, van Griensven A, et al. Comparison of the performance of six drought indices in characterizing historical drought for the upper Blue Nile basin, Ethiopia. Geosciences. 2018;**8**:81

[18] Zeleke TT. Trend and periodicity of drought over Ethiopia. International Journal of Climatology. 2017. DOI: 10.1002/joc.5122

[19] Viste E, Korecha D, Sorteberg A. Recent drought and precipitation tendencies in Ethiopia. Theoretical and Applied Climatology. 2013. DOI: 10.1007/s00704-012-0746-3

[20] Teshome A. Increase of extreme drought over Ethiopia under climate warming. Hindawi: Advances in Meteorology. 2019;**2019**:18. DOI: 10.1155/2019/5235429

[21] Ali A. Economics of Resilience to Drought—Somalia Analysis. USA: USAID; 2017. pp. 1-45

[22] Philip S, Kew SF, van Oldenborgh GJ, Otto F, O'Keefe S, Haustein K, et al. Attribution analysis of the Ethiopian drought of 2015. Journal of Climate. 2018;**31**:2465-2486

[23] Chemeda D, Mukand E, Babel S. Drought analysis in the Awash river basin, Ethiopia. Water Resources Management. 2010;**24**:1441-1460

[24] Gebrehiwot T, van der Veen A, Maathuis B. Spatial and temporal assessment of drought in the Northern highlands of Ethiopia. International Journal of Applied Earth Observation and Geoinformation. 2011;**13**:309-321

[25] Gissila T, Black E, Grimes DIF, Slingo JM. Seasonal forecasting of the Ethiopian summer rains. International Journal of Climatology. 2004;**24**:1345-1358

[26] Awass AA. Hydrological Drought Analysis—Occurrence, Severity, Risks: The Case of Wabi Shebele River Basin. German: University of Siegen; 2009 [Dissertation]

[27] Van Lanen HAJ, Enyew B. Assessment of the impact of climate change on hydrological drought in Lake Tana catchment, Blue Nile basin, Ethiopia. Journal of Geosciences. 2014;**03**:174

[28] Berhanu B, Seleshi Y, Melesse AM. Nile River Basin: Ecohydrological Challenges, Climate Change and Hydropolitics. Switzerland: Springer International Publishing. pp. 1-718

[29] Bayissa YA, Moges SA, Xuan Y, Van Andel SJ, Maskey S, Solomatine DP, et al. Influence de la durée des chroniques météorologiques sur la caractérisation des sécheresses météorologiques du bassin supérieur du Nil Bleu (Ethiopie). Hydrological Sciences Journal. 2015;**60**: 1927-1942

[30] Masih I, Maskey S, Trambauer P. A Review of Droughts in the African Continent: A Geospatial and Long-Term Perspective. European Geosciences Union. Hydrology and Earth System Sciences. Copernicus Publications; 2014

Section 2

Climate Change and Drought

Chapter 3

Climate Change: A Real Danger to Human and Animal Survival

Godwill Makunde, Nation Chikumba, Walter Svinurai and Xavier Mhike

Abstract

Some countries in Southern Africa where hit by either a storm or cyclone or both in 2019 alone manifesting a changing climate. Infrastructure and cropping land was destroyed, both animal and human lives were lost due to the flooding events. Drought is a common phenomenon in this region, often occurring once in three years. This has affected food, feed and nutritional security of both humans and livestock. Saline soils unsuitable for agriculture, other animal and plant life are expanding fast due to insufficient precipitation. Soil degradation is on the rise, leaving soils with poor water holding capacity to support sustainable agriculture. Climate change is changing the environment and new pests and diseases for both crops and livestock are emerging. World governments, industries and general populace should find better ways of reducing air pollution by greenhouse gases which have a net effect of damaging the ozone layer and increasing atmospheric temperatures. At the same time, plant and animal breeding should aim at improving crop cultivars and animal breeds that resist to the constraints such as drought and heat stress brought by climate change. The human population is increasing at an alarming rate and need both food and nutritional security.

Keywords: drought, floods, heat stress, animal diseases, plant pests

1. Introduction

The whole earth's climate systems are changing in the atmosphere, the oceans, ice floes and on the land. Some of the changes such as increases in temperature, rise in carbon dioxide, drought and floods are already in motion now, while others like continued sea level rise are already irreversible for centuries to millennia ahead. The impact of climate change makes lives of both humans, animals, and plants unbearable. The main drivers of climate change are attributed on human activities in the following economic sectors, transport (road, air, rail, and sea), energy industries (electricity, heat, power) and agriculture [1]. Agriculture alone contributes to almost 32% of all greenhouse gases that contribute to climate change. Human population is increasing in the world and assuring an increase in demand for food, shelter, water, and clothing.

Africa is one of the world's most vulnerable regions due to the fragility of its economies. It is now evident that global warming in the 21st century will be more intense in Africa compared to the rest of the world [1, 2], Global warming and increased climate variability will severely affect crop and livestock production systems in Africa. Some of the harsh realities of climate change has been felt in Southern Africa in recent years but the worst scenarios of Storm Desmond (January 23, 2019, in central Mozambique), Cyclone Idai (March 14, 2019, in central Mozambique, Zimbabwe, Malawi and northern Madagascar) (Cyclone Idai: Wikipedia) and cyclone Kenneth (April 25, 2019). A total of 100,000 homes were destroyed and more than 1000 people were killed during the cyclones and storm. **Figure 1**. shows the areas that were affected by cyclone Idai in southern Africa while **Figure 2**. shows some of the damage done in Mozambique.

Villages, towns, infrastructure, cropped land were destroyed, and livestock died during the cyclone events. One in every five years, agricultural and ecological drought often affects most countries in southern Africa, sometimes leading to 100% crop failure depending on timing, intensity, and severity. Definition of drought varies with for different water users. Meteorological, Hydrological and Agricultural drought is a prolonged period with shortage of precipitation/below average precipitation, surface water or soil moisture. Climate change effects develop slow over time and their impact underestimated—negative impact on vegetation, animals, and people. Livestock and wild animals also suffer from crop and veld failure. The natural production of water (water cycle) is intensified by climate change and brings intense rainfall and associated flooding and intense drought in some regions. The rainfall patterns and distribution are affected. Significant economic (loss of employment, decreased agricultural and industrial production) and social disasters, such as famine, forced migration, and conflict over few remaining resources, health related- lack of water, poor nutrition and famine are some of the negative effects of drought.

Figure 1.
Areas in southern Africa which were affected by cyclone Idai in march 2019, destroying infrastructure and crops as well as killing people and animals (source: Cyclone Idai: Wikipedia).

Figure 2.
Flooded villages and destroyed crop lands in Mozambique in 2019 (source: Cyclone Idai: Wikipedia).

Climate change causes warming up of the earth resulting in severe heat waves in some regions. Hotter days by 6 to 15°C are expected in some countries while other will suffer colder days. All the changes in temperature and environment give rise to new pests and diseases for both humans and animals. Serious crop and animal yield losses to emerging pests and diseases will lead to starving populations.

Climate change affects prediction of the future and is an indication that the past is over. It is still difficult to assess the extent and nature of such changes in the future.

2. Agriculture and climate change

Agriculture in its various forms anchors the core component of human life. Vital is crop and livestock production to the agricultural sector. Worldwide, agriculture is the largest economic sector and in Africa as well as Asia employs between 70 and 90% of the total labour force as well as supplying up to 50% of household food requirements and household incomes [3].

2.1 Livestock and crop production in Africa

Livestock production contributes between 30 and 50% of agricultural gross domestic production, mainly from the production of beef cattle, dairy cattle, goats, sheep, and chickens [4]. Africa has about 250 million Tropical Livestock Unit (TLU) equivalents [3] that support 70% of the rural livelihoods and provide income to over 200 million people. Livestock production is increasing throughout Africa, due to rising human population, urbanisation, and demand for meat. Total human population in the world is expected to reach 8.6 billion by 2035 and 9.8 billion by 2050 [5]. In addition to provision of food security and income, livestock provide draught power, transport, and manure in mixed crop-livestock systems. Also, livestock is socially and culturally important for payment of dowry, celebrations, and gifts to family members, and as a means of savings [6].

The livestock sector is widely constrained by a lack of regulation, which leads to negative externalities such as land degradation, water pollution, loss of biodiversity and emission of greenhouse gases (GHGs). Greenhouse gasses are the major causes of climate change, and the consensus is that climate change and global warming are real due to increased GHG emissions into the atmosphere.

2.2 Impact of climate change on livestock and crop/forage yields and quality

Livestock production depends on natural resources, which in much of Africa, primarily means pasture and water [7, 8]. The transition towards drier and hotter regimes with climate change is expected to have several adverse impacts on the quantity and quality of animal feeds. Grazing areas will be affected by changes in herbage growth brought about by changes in atmospheric carbon dioxide concentrations and temperature [9]. In terms of field crops, climate change has affected yield in several places. Some crop yields have increased in some places and decreased in others. Yield of some crops such sorghum has increased by 0.7% in sub-Saharan Africa and 0.9% in Asia due to favourable environments created by climate change [10]. In other instances, yield of some world staples such as rice and wheat are reducing on average by 0.3 and 0.9% respectively. In some studies, climate change was reducing food calories by 1% for the top 10 global crops. The food calory reduction is happening throughout the world, both in rich industrialised countries and poor countries.

2.3 Effects of heat stress

Heat stress is defined as the rise in air temperature above a threshold level for a significant amount of time to cause damage to normal growth and development [11]. Heat stress limit plant growth and productivity even in the presence of adequate soil moisture. Just like drought, heat stress can happen at any stage of crop growth and its negative effect varies with the onset, intensity, and duration during plant growth. Field crops which give flowers, pollen, tassels, silking, grain filling, storage root formation and bulking are most susceptible to heat stress at reproductive stages due to flower drop. Yield losses can reach 100 percent depending on intensity and duration of the heat. In addition, an increase in air temperature results in raised soil temperatures which can be higher than air temperature when soil moisture is limited. Root development is severely affected in both field crops and forage/pastures for animals.

The vulnerability of livestock to heat stress varies according to species, genetic potential, life stage and nutritional status. The projected warmer temperatures expected in sub-Saharan Africa are likely to cause heat stress in beef cattle raised under extensive production systems. Heat stress in extensive beef cattle production systems will likely reduce foraging time, feed intake, growth performance and carcass quality. Reproductive performance will also be compromised, for example, conception rates will decrease, calving intervals will increase, and spermatogenesis and semen quality will be impaired. Heat-induced reduction in feed intake will result in a decline in milk yields in dairy cattle [12].

The effect of heat stress is not likely to be as adverse in small ruminants, due to the small body weight, well developed water retention in the kidney and lower metabolic rates of smaller ruminant species. Goats are more likely to cope with, and adapt to, the increasingly hot and dry conditions expected in the region compared to sheep and cattle, because of their low feed and water requirements, ability to exploit low quality forage and disease resistance. Goats can survive harsher climates than cattle and require less space [12].

2.4 Availability of water and its usage

Climate change will amplify existing stress on water availability in agricultural systems of semi-arid environments [13]. Rising temperatures may increase irrigation water requirements of major crops [14] and drive-up water demand by livestock [9, 15]. For example, the increased reliance on groundwater in the future in Botswana for the cattle sector could lead to problems associated with the sustainability of water resources in the country [16]. Global warming and accompanying hydrological changes are also likely to affect soils in complex ways, including soil fertility and propensity for erosion [14]. Additionally, much prime agricultural land located in the coastal plains of Southern Africa might be lost to rising sea-levels [17]. Most of staple foods consumed in sub-Saharan Africa are grown under rain fed conditions largely by small scale farmers who have limited capacities to effect supplementary irrigation in cases of drought. Southern Africa and West Africa have a one rainfall season in a year and suffer drought episodes.

2.5 Disease and pest outbreaks on livestock and crops

Several studies have shown that the impact of climate change on the transmission and geographical distribution of animal diseases will vary according to the ecosystem, the type of land use, disease-specific transmission dynamics, susceptibility of the populations at risk and sensitivity of the pathogen to temperature and humidity [3, 18]. Climate change is expected to alter transmission rates between hosts by affecting the survival of the pathogens or parasites or the intermediate vectors, but also by other, indirect, forces that may be hard to predict with accuracy. For example, a series of droughts in East Africa between 1993 and 1997 resulted in pastoral communities moving their cattle to graze in areas normally reserved for wildlife. This resulted in cattle infected with a mild lineage of rinderpest transmitting disease both to other cattle and to susceptible wildlife such as buffalo and impala, causing severe disease, and devastating certain populations [19].

Climate change is also expected to affect the abundance or distribution of hosts or the predators of vectors and influence patterns of disease in ways that cannot be predicted from the direct effects of climate change alone [20]. Climate change-related disturbances of ecological relationships, driven perhaps by agricultural changes, overgrazing, deforestation, construction of dams and loss of biodiversity, could give rise to new mixtures of different species/strains, thereby exposing hosts to novel pathogens and vectors and causing the emergence of new diseases [21].

2.5.1 Influence on pathogen diversity and virulence

Higher temperatures may influence some pathogens and parasites through accelerated development on their life cycle outside their hosts. Pathogens and parasites that are sensitive to moist or dry conditions may be affected by changes to precipitation, soil moisture and the frequency of floods [20]. Changing wind patterns could affect the spread of certain pathogens and vectors, particularly the infective spores of anthrax and blackleg, the wind-borne dermatophilosis [22].

2.5.2 Influence on animal and crop hosts

Climate change may bring about substantial shifts in diseases distribution, and outbreaks of severe diseases could occur in previously unexposed animal and plant populations. While livestock often have evolved genetic resistance to diseases to which they are commonly exposed, they may be highly susceptible to 'new diseases' [3]. For example, mammalian cell immunity level may be suppressed following a sharp exposure of light of ultraviolet B nature because of expected ozone layer depletion [23]. Ultraviolet B depletes specific lymphocyte cells and animals become susceptible to some pathogens such as viruses; rickettsia (such as *Cowdria* and *Anaplasma*, and some bacteria, such as *Brucella* [18]. Continued depletion of ozone layer would, therefore, possibly impact some animal diseases in future. Endemic stability of animals is likely to be disrupted for tick borne diseases such as anaplasmosis, babesiosis and cowdriosis that exhibit endemic stability [24]. If climate change drives such diseases to new areas, non-immune individuals of all ages in these regions will be newly exposed, and outbreaks of severe disease could follow [18].

2.5.3 Influence on disease vectors

Changes in moisture and temperature regimes may impose limits on the distribution and the abundance of vectors. Often, low temperatures limit vector distribution because of high winter mortality and a relatively slow rate of population recovery during warmer seasons [20]. This is different with high temperatures as, limiting occurs when there is excessive moisture loss. Therefore, cooler, and high-altitude regions which were previously too cold for certain vectors may begin to allow them to flourish with climate change. Warmer regions could become even warmer and yet remain permissive for vectors if there is also increased precipitation or humidity. For example, biting midges and mosquito-borne diseases outbreaks have been linked to the occurrence of ENSO [25–27].

2.5.4 Additional effects of climate change on animal and crop productivity

Climate change, in conjunction with other forms of land use fosters the emergence of new diseases as it changes the structure of the ecosystems in relation to species composition and diversity in favour of livestock pathogens and vectors [28]. Similarly, the preponderance of diseases due to climate change occurs through tendencies of animals to migrate in masses, overgraze and congest around pastures, during times of droughts, a common phenomenon during this climate change era [3]. High rains on the other hand positively correlate with disease outbreaks in both livestock and crops. In field crops, diversity will be lost, and pathogens will rely on a few crop species, leading to increases in diseases such as leaf spots, root rots, blights, and cankers. In livestock production, apart from increases in internal and external parasites, diseases like dermatophytosis, anthrax and foot rot will rapidly occur and spread. This way, areas with limited disease occurrences may end up experiencing high crop and livestock disease occurrences. Changing international trade patterns, local animal and crop transportation, farm size and human migrations are all factors that may be driven in part by climate change, and which may impact negatively on disease transmission.

3. Strategies for adaptation to climate change in the livestock sector

For long, there have been widely publicised calls to mitigate and adapt to climate change issues and efforts are evident worldwide, however, this may be at the expense of natural ways of adaption and fears are that there could be the possibility of the emergency of new unforeseen challenges on the environment [3]. Adaptation typically refers to longer-term changes in behaviour and practices which are more likely to reduce underlying vulnerability to climate change [29]. Commonly documented livestock adaptation strategies include diversifying livestock activities, diversifying livestock types, supplementing livestock feed, and developing niche markets to preserve indigenous breeds.

3.1 Niche breeding of locally adapted animals

Indigenous livestock breeds have an adaptive to their native localities advantage than the exotic ones. Characteristics such of drought resistance, disease tolerance are crucial for sustainability of livestock in the face of climate change and indigenous breeds are inherently superior in this regard. In Africa, there are higher levels of inbreeding of local breeds and these animals are often found in small-scale and pastoralist farming setups. Apart from having an adaptive advantage, knowledge of the ways of survival of such breeds ensures an effective way of maintaining the natural environments in which they dwell in with limited chances of negatively affecting them. The success of pasture and rangeland conservation lies in part on the availability of livestock breeds that can effectively utilise the environment [30].

3.2 Animal diversification at farm level

Diversification of animal species provides a means through which a broad range of plant species may be utilised. As climate change influences temporal and spatial variations in the vegetation nature with respect to species diversity, quantity and quality of the biomass and having a diversified herd of animals may help to close this gap. Places rich in shrubs for example may be better suited for goats while those with grasses may be well suited for cattle and diversification in this scenario offers complete utilisation of the available vegetation [31].

3.3 Adopting livestock production as an adaptation strategy

The integration of livestock farming with crop production, where some of the crops grown at the farm serve as livestock feed has been a common practice across the globe. This however in most cases depends is a function of matching the correct livestock type with the feed available, the ability of the animal in question to survive on crop residues or disease resistance.

Poultry production has since been one of the enterprises engaged by many farmers in collaboration with crop production in many countries, [32]. Poultry can effectively utilise crop-based feeds such as soya beans and maize with high production success rates. Many indigenous chicken breeds can even survive under free range production systems with little or no feed supplementation. Similarly, locally adapted breeds have high disease resistance rates corresponding to a limited need for prophylactic and therapeutic disease management. Poultry are particularly especially in Africa a means of controlling the effects of climate change as they require minimal production space and can thrive on household waste.

3.4 Reviving and creation of markets for indigenous breeds

The maintenance of local adapted breeds plays a pivotal role to produce breeds adapted to the changes in climate in future. These advantages are however now being eroded by the rising need by breeders to produce breeds with high performing traits, especially characteristics of economic importance. About 11 and 2% of mammalian and avian breeds, respectively have been reported to be extinct in recent years with a further 210 cattle breeds and 179 sheep breeds classified under the critically endangered species [33]. Maintaining local breeds requires a multi-pronged approach, starting with respect for the rights of local custodians of these breeds and support for their production systems.

The quest of maintaining local breeds is inevitable in the face of climate change however for these breeds to be competitive, niche markets through improving production; processing and value addition of their products is an area of prime focus. Such markets are already in existence although there is still much to be done to smoothen the intervention. Performing SWOT analysis is an important strategy for driving the evolution towards identification of promising niche markets. The most important domain of the process is to organise local people who have the breeds of interest and ensuring that they are directly involved in the process.

4. Adaptation strategies in the crops sector to climate change

Apart from agronomic measures such as water harvesting, irrigation, mulching and growing of well adapted crops and varieties, plant breeding offers more sustainable and widespread answers to a changing climate. Plant breeding goals should meet the changing climate. Major goals should include the following:

4.1 Minimization of postharvest losses

Plant breeding programs should include postharvest durability in their product profiles. Huge losses of between 40 and 70% of cereals and legumes harvested in the world are lost to grain pests mainly weevils and moths. In fruit and vegetable sector, 25–50% of the produce is lost at postharvest stage, while 20–40% of cassava is lost to post physiological deterioration and sweetpotato to moulds. The control or management of postharvest losses is the easiest way to double food availability in Africa. With increased temperatures due to climate change, anticipated losses could be higher than mentioned.

4.2 Heat tolerance

Heat stress during reproductive, grain filling, storage root formation and root bulking leads to severe yield losses in field crops. Heat stress has damaging effects irrespective availability of other factors required for normal plant growth. Breeding for heat tolerance in field crops is one of the strategies of coping with high temperatures induced by climate change. Climate model predictions and simulations pointed to an increase between 1.8 and 4 degrees Celsius in temperature in the 21st century with high increases in southern Africa and Southeast Asia [2].

4.3 Drought tolerance

Climate change alters the precipitation amount, patterns, and distribution. The occurrence of drought and its severity has already been felt in some regions of the world and is expected to significantly rise soon. Food and nutrition security is threatened by frequent droughts in Southern Africa. The release of cultivars with improved tolerance to drought will reduce the impact of drought on many nations in southern Africa whose economies are already struggling from COVID-19 pandemic and political disturbances. Many methods and techniques are available for breeding programs to utilise in the development of drought tolerant cultivars. Genomic selection tools and emergence of different bodies ready to capacitate breeding programs in modernization and operational excellence would help in the identification of drought tolerant genotypes for respective. Water use efficiency is one mechanism for drought tolerance which should be explored in breeding programs.

4.4 Salinity tolerance

The earth is receiving lesser rainfall than before and experiencing more evaporation making it impossible to dilute salts in the soil. In addition, sea encroachment is recorded due to rising seas and is depositing a lot of salt in agricultural land. Breeding for salt tolerance should be considered high priority in nations receiving less precipitation duc to climatc changc.

4.5 Tolerance to low soil fertility especially nitrogen and phosphorus requirements

Soils are continuously depleted and there is higher need for crops with high efficiency in the use of nitrogen and phosphorus. Legume crops with high capacity of nitrogen fixation bred and released for commercialization should be compatible with free living N fixing bacteria, Rhizobia. This helps restore nitrogen in the soils which could benefit staple cereal crops in subsequent crop rotations.

4.6 Pest and disease resistance

Climate change brings in new pest and diseases in a similar manner to livestock. Plant breeding need to continuously match the evolving pests and diseases that would reduce crop productivity and produce quality.

4.7 Recommendations

Communities as well as nations are likely going to fight for water resources, pastures, and other few comfortable places to live due to diminishing natural resources particular rainfall and water. There is an on-going tension between Egypt and Ethiopia over water resources from Nile River. Conflict resolution skills and techniques should be endowed in leaders at all levels and should be taken seriously.

Focussed niche breeding especially for targeted production environments should be encouraged in both animal and crop breeding programs. Global campaigns to reduce emissions of greenhouse gases should be led by governments and multilateral companies. Innovative ways of food and feed conservation should be developed and adapted to reduce post-harvest losses.

5. Conclusions

Climate change is a reality and solutions to reduce its effect on humans, animal and the environment should be put in place and be adapted by all nations. In summary, Africa's agricultural sector and human population will suffer the effects of climate change in a multitude of ways:

- frequent droughts caused by reduced precipitation and shifts in growing seasons particularly in Southern Africa.
- occasional floods caused by too much rainfall with short periods of time. This is witnessed more in Southern Africa most affected countries being Madagascar, Mozambique, Zimbabwe, and Malawi.
- rise in temperatures and becoming uncomfortable for life for both humans and animals. Crop and animal growth will be affected severely reducing productivity and production.
- forced migrations by both animals and humans. This will exert a lot of pressure on habitable environments and economies.
- emergence of new diseases and pests to both humans and animals which might be difficult to cure.
- the cost of food and feed will increase due to decreased productivity and production in certain environments.

DOI: http://dx.doi.org/10.5772/intechopen.103022

Author details

Godwill Makunde[1*], Nation Chikumba[2], Walter Svinurai[2] and Xavier Mhike[3]

1 International Potato Center, Maputo, Mozambique

2 Marondera University of Agricultural Sciences and Technology, Marondera, Zimbabwe

3 Agricultural Research Council of Zimbabwe, Harare, Zimbabwe

*Address all correspondence to: g.makunde@cgiar.org

References

[1] IPCC. Global Warming of 1.5°C. An IPCC Special Report on the Impacts of Global Warming of 1.5°C above Pre-Industrial Levels and Related Global Greenhouse Gas Emission Pathways, in the Context of Strengthening the Global Response to the Threat of Climate Change, Sustainable Development, and Efforts to Eradicate Poverty. In: Masson-Delmotte V et al., editors. Geneva, Switzerland: World Meteorological Organization; 2018

[2] IPCC. Climate Change 2007: Impacts, Adaptation and Vulnerability. Contribution of Working Group II to the Fourth Assessment Report of the Intergovernmental Panel on Climate Change. Cambridge, UK: Cambridge University Press; 2007

[3] International Union for Conservation of Nature (IUCN). 2010. The Nature of Progress: Annual Report 2010

[4] IFAD. Livestock and climate change. In: Draft Thematic Paper. January 12-13, 2009. Rome, Italy; 2009

[5] The World Population Prospects: The 2017 Revision. www.un.org/en/desa/world-population-prospects-2017-revision

[6] World Bank. The World Bank Annual Report 2009: Year in Review. Vol. 1. Washington, DC: World Bank; 2009

[7] Henry B, Charmley E, Eckard R, Gaughan JB, Hegarty R. Livestock production in a changing climate: Adaptation and mitigation research in Australia. Crop & Pasture Science. 2012;**63**:191-202

[8] Polley HW, Briske DD, Morgan JA, Wolter K, Bailey DW, Brown JR. Climate change and North American rangelands: Trends, projections, and implications. Rangeland Ecology Management. 2013;**66**:493-511

[9] Silvestri S, Bryan E, Ringler C, Herrero M, Okoba B. Climate change perception and adaptation of agro-pastoral communities in Kenya. Regional Environmental Change. 2012;**12**(4):791-802

[10] Ray D. Climate change is affecting crop yields and reducing global food supplies. 2019. Available from: https://allianceforscience.cornell.edu/blog/2019/07/

[11] Wahid A, Gelani S, Ashraf M, Foolad MR. Heat tolerance in plants: An overview. Environmental and Experimental Botany. 2007;**61**:199-223

[12] Dzama K. Is the livestock sector in Southern Africa prepared for climate change? Southern African Institute of International Affairs (SAIIA) Policy Briefing 153. 2016

[13] Field CB, Barros VR, Mach KJ, Mastrandrea MD, van Aalst M, Adger WN, et al. Technical summary. In: Climate Change 2014: Impacts, Adaptation, and Vulnerability. Contribution of Working Group II to the Fifth Assessment Report of the Intergovernmental Panel on Climate Change. UK and New York, USA: Cambridge University Press; 2014

[14] Morton JF. The impact of climate change on smallholder and subsistence agriculture. 2007

[15] Thornton PK, Van de Steeg J, Notenbaert A, Herrero M. The impacts of climate change on livestock and livestock

systems in developing countries: A review of what we know and what we need to know. Agricultural Systems. 2009;**101**(3):113-127

[16] Masike S. The impacts of climate change on cattle water demand and supply in Khurutshe, Botswana [PhD thesis]. Hamilton, New Zealand: University of Waikato; 2007

[17] Chishakwe NE. Southern Africa Sub-Regional Framework on Climate Change Programmes Report. 1st Draft Working Document. SADC–UNEP. 2010

[18] Baylis M, Githeko AK. The Effects of Climate Change on Infectious Diseases of Animals foresight report office of science and innovation; 2002. pp. 24-37

[19] Kock RA, Wambua JM, Mwanzia J, Wamwayi H, Ndungu EK, Barrett T, et al. Rinderpest epidemic in wild ruminants in Kenya 1993-1997. Veterinary Record. 1999;**145**:275-283

[20] Kimaro EG, Chibinga OC. Potential impact of climate change on livestock production and health in East Africa: A review. Livestock Research and Rural Development. 2013;**25**(7):116

[21] WHO. Climate Change and Human Health. World Health Organisation: Geneva Article. Retrieved on 23 April 2013. 1996. Available from: http://www.who.int/globalchange/publications/climchange.pdf

[22] Thornton PK, Jones PG, Owiyo T, Kruska R, Herrero M, Kristjanson PM, et al. Mapping climate vulnerability and poverty in Africa. International Livestock Research Institute, Nairobi, Kenya. Africa Journal of Agriculture and Resource Economics. 2008a;**2**:24-44

[23] de Gruijl FR, Longstreth J, Norval M, Cullen AP, Slaper H, Kripke ML, et al. Health effects from stratospheric ozone depletion and interactions with climate change. Photochemical and Photobiological Sciences. 2003;**2**:16-28

[24] Eisler MC, Torr SJ, Coleman PG, Machila N, Morton JF. Integrated control of vector-borne diseases of livestock: Pyrethroids: Panacea or poison? Trends in Parasitology. 2003;**19**:341-345

[25] Anyamba A, Linthicum KJ, Mahoney R, Tucker CJ, Kelley PW. Mapping potential risk of Rift Valley fever outbreaks in African savannas using vegetation index time series data. Photogrammetric Engineering and Remote Sensing. 2002;**68**:137-145

[26] Baylis M, Mellor PS, Meiswinkel R. Horse sickness and ENSO in South Africa. Nature. 1999;**397**:574

[27] Gagnon AS, Smoyer-Tomic KE, Bush ABG. The El Nino Southern Oscillation and malaria epidemics in South America. International Journal of Biometeorology. 2002;**46**:81-89

[28] Nardone A, Ronchi B, Lacetera N, Ranieri MS, Bernabucci U. Effects of climate change on animal production and sustainability of livestock systems. Livestock Science. 2010;**130**:57-69

[29] Vincent K, Cull T, Chanika D, Hamazakaza P, Joubert A, Macome E, et al. Farmers' responses to climate variability and change in southern Africa–is it coping or adaptation? Climate and Development. 2013;**5**(3):194-205

[30] World Initiative for Sustainable Pastoralism (WISP). 2008b. Policy Note 8. Forgotten Services, Diminished Goods: understanding the agroecosystem of pastoralism

[31] Behnke R, Scoones I, Kerven C. Range Ecology at Disequilibrium: New

Models of Natural Variability and Pastoral Adaptation in African Savannas. London: Overseas Development Institute; 1993

[32] Alders R, Bagnol B, Harun M, Young M. Village poultry, food security and HIV/AIDS mitigation. LEISA. 2007;**3**:20-21

[33] FAO. Global Action Plan for Animal Genetic Resources. Interlaken, Switzerland, 3-7 September; 2007

Chapter 4

Reducing the Effects of Drought and Degradation of Agricultural Soils, in the Context of Climate Change, through the Application of Regenerative Ecological Technologies

Eugen Popescu, Florin Nenciu and Valentin Nicolae Vladut

Abstract

The agricultural sector has a limited capacity for expansion, consequently, deficient technologies based on the widespread use of synthetic chemicals have been implemented in the last decades, having a major negative impact on natural ecosystems, biodiversity, and environmental services. Desertification, land degradation, and drought, combined with human activity and environmental changes, cause important soil losses and a reduction in natural defenses against droughts and floods. The combined impact of climate change, land mismanagement and unsustainable freshwater use has long been affecting agricultural productivity, the most common cause being unsustainable land management practices. This chapter aims to briefly assess the most effective strategies for reducing the impact of climate change on agricultural crops, as well as to prevent or reverse the process of desertification and systematic loss in food quality and quantity. Regenerative management practices such as minimum tillage technologies, cover crops and mulching, inoculation with microorganisms, nutrients cycling, the balance of the organic fertilizers or foliar application help farmers in managing healthy soils, capable of growing rich and ecological crops without the use of chemical hazardous substances.

Keywords: climate change, regenerative ecological technologies, desertification, land degradation and drought

1. Introduction

Regenerative agriculture is a farming and land management concept based on several principles and techniques that strengthen and restore ecosystem functions and health. Long-term usage of regenerative agriculture has shown many benefits in terms of quality and profitability for farmers, as well as improving the environment and contributing to the maintenance of a healthy agricultural landscape. Given that it is not always very clear how each action contributes to better agricultural management and drought mitigation, this chapter aims to recall the key elements that farmers must consider in regenerative agriculture in order to have the best results. It should be noted that there is more than one approach that may differ depending on local circumstances, however, the elements described in this chapter serve as a starting point for practitioners and academics who wish to learn more or deepen one of the related domains.

The existence of life is largely dependent on the richness and health of soils, which is why soil structure, together with water availability are the most valuable resources for humanity. The annual degradation of the agricultural lands puts even more pressure on farmers, forcing them to use more chemical inputs and these practices may eventually lead to extreme phenomena such as drought, floods, and eventually soil abandonment [1]. However, both farmers and policymakers continue to be neglecting the need for soil health preservation and they do not take firm restoration measures even when the situation becomes concerning.

Water, minerals, and organic matter combine to make the soil in a natural process. Soil minerals are produced in the process of natural erosion, while the organic matter is formed by the decomposition of plants and other organisms that have died. Many scientists consider soiling a finite resource that cannot be renewed during a human lifespan. We propose, in the present chapter, several techniques used and validated for faster restoration of soil properties, which may help recovery in very shorter time periods, depending on the degree of soil impairment.

Degraded soil is described as a change in the physical, chemical, and biological characteristics that results in a reduced capacity to support plant growth. The most common phenomena that usually occur are related to the fact that soil loses the capacity to deliver nutrients and water, while toxic compounds restrict plant growth, topsoil lacks of organic matter content, subsoil resources are insufficient to support plant roots, the compaction rate is substantially increased, drainage occurs with difficulty, and many of the needed microorganisms are absent.

In most common cases, the quality of the soil decreases as a result of the anthropogenic intervention, while some natural causes are aggravating the circumstances, often leading to erosion. Human activity is the most frequent cause of agricultural soil degradation and for accelerating natural soil erosion. Agriculture has deteriorated the Earth's soils during the last 100 years, with disastrous consequences, David R. Montgomery [2] estimates that humanity is losing 0.3% of our global food production each year due to soil erosion and degradation. Soil degradation and loss has been a problem since the beginning of agriculture and played a major role in the demise of past civilization including Mesopotamia, Antic Greece, and the Rome Empire. The element that contributes probably the most to the negative damage to the soil, more important even than deforestation is the plowing activity. Stanford University in a study from 2015 estimated the degradation rate of topsoil worldwide at a rate of 70%, with margins between 54% in Africa and 74% in North America [3]. At this time, there is no allotted restoration period, since we are eroding soil 20 times faster than we are regenerating it.

Degraded soils have a poor health state, reducing the ecosystem's ability to provide water and nutrients to plants, and affecting the soil nutrient web. Degraded soils have a weak structure attributable to a lack of soil biodiversity, which causes flooding, erosion, and low production. Water cannot penetrate inferior soil structures, so the rains follow the flow of gravity, transporting major amounts of minerals and salts to the groundwater, rivers, or lakes. During a drought period, there will be no moisture, and groundwater will not be replenished easily. Plants will be stressed, and yields will decrease very fast. In the tropics especially on fertile land, soil deterioration is prevalent. Natural erosion caused by wind, sun, severe rainfall, and poor human management are the most common causes.

It is critical to understand that poor agricultural management before and during a drought has a synergetic effect on soil properties. Land degradation in arid, semiarid, and sub-humid areas may be generated by various external factors including climate change, and draught may lead to desertification. Desertification may be irreversible if not intervened in time, especially when the environment becomes too dry and the soil becomes further degraded through erosion and compaction.

One of the most important hazardous environmental events in recent history was the American Dust Bowl during the years 1930–1936, when large dust storms swept topsoil from significant land areas, making 75% of the original topsoil quality to be lost [4]. Commenting on the American Dust Bowl, US President Franklin D. Roosevelt said "The nation that destroys its soil, destroys itself", a remark that is still relevant to modern crop management practices.

Storms, torrential rains, floods, and droughts are becoming more frequent and intense as a result of climate change. Every year, soil deterioration worsens, plants get stressed, and yields decrease. Soil management is, therefore, an essential element of sustainable agriculture.

Proper regenerative soil management will slow down or stop soil degradation and start to rebuild soil fertility. Management should be focused on obtaining healthy and superior plants that do not need intense chemical treatments sincc it is proven that a high immunity system protects crops from diseases and insect attacks. Increasing plant immunity will be pointed out in high yields, quality products, plants will get increased resistance to diseases and pest attacks. At the same time soil will become healthier, full of nutrients with an active and rich soil food web. Healthy soils with a balanced nutrients ratio, promote biological high activity and replenish groundwater, and will help the plants to withstand better the drought. To stop soil degradation, special attention must be paid to the phenomena that produce natural erosion, and rejuvenate the soil, while human activities have to change rapidly. Soil regeneration practices sequester an important part of the required amount of carbon in the soil, allowing mankind to maintain control over climate change. Soil carbon allows the land structure to function as a sponge, each gram of carbon-absorbing 8 grams of water. In addition to the positive effect on the mineralization process, carbon helps to build the soil structure, which aids in the supply of air, water, and nutrients to plants. Plants, in response, emit liquid carbon from their roots, increasing, even more, the synergies and water absorption. This phenomenon occurs more frequently when aggressive tilling works are avoided, and the synthetic fertilizer and synthetic biocides application are not used. The techniques, if they are applied indiscriminately have the opposite result, eliminating the carbon. **Figure 1** depicts the most common approach to regenerative agriculture at three levels of management: acknowledge the objectives and benefits, comprehend the fundamental concepts, and put the best practices into action.

Regenerative agriculture requires a complete redesign of the farming system, as well as a shift in the procedures and metrics used in traditional agriculture, and a longer-term commitment of farmers.

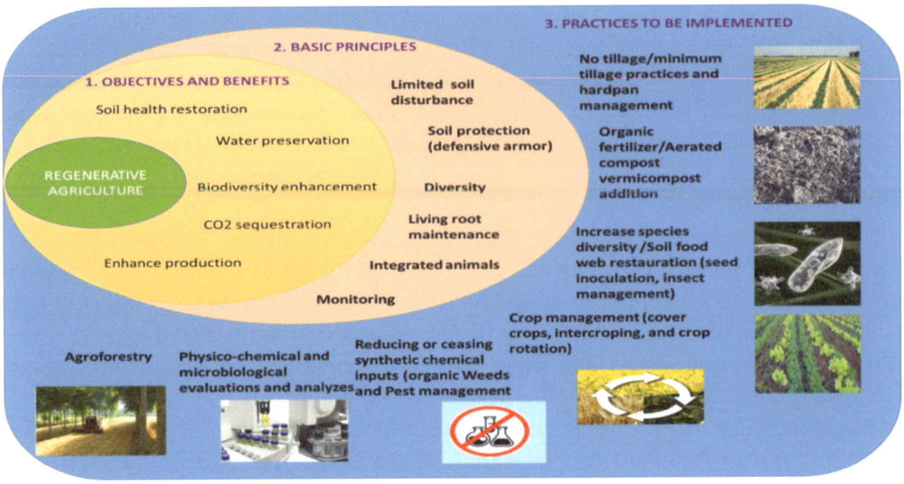

Figure 1.
A simplified approach to regenerative agriculture implementation.

2. Main causes of soil degradation

FAO [5] defines soil degradation as a change in the soil health status, resulting in a diminished capacity of the ecosystem to provide goods and services for its beneficiaries. Degraded soils have a state of health that prevents them from generating the standard products and services in a given ecosystem. Soil degradation is caused by unfavorable interaction between physical, biological, and chemical soil characteristics, accelerating erosion, and leading to poor drainage, salinization, nutrient imbalance, decrease in soil organic matter, and suppressing biology. Physical soil deterioration includes changes in soil structure (crusting, compaction, etc.), imbalance in water content and air ratio, leading to extreme surface temperatures. Chemical soil deterioration includes nutrient leaching, fertility depletion, or even toxicity. Biological deterioration includes a decrease in the microorganism population and a drop in their activity, as well as, a severe reduction of organic matter content. Degraded soil is being studied at specialist institutions in nearly every country, and warnings are coming from all across the scientific world [6–8].

Major causes of soil degradation are divided into natural, as climate variations (soil degradation caused by wind, sun, drought, or heavy rains favoring the fertile soil to be washed away) and anthropogenic activities (overgrazing, deforestation, excessive use of chemicals fertilizers, pesticides, herbicides, bare soils, excess of tillage, overdraft of groundwater, etc.) [9].

Conventional agriculture is considered to be one of the biggest contributors to soil degradation [10]. After Second World War, the Chemical Industry provide agriculture with new and advanced chemical formulas used as fertilizers, herbicides, and pesticides. The first results showed great success for everyone; however, the long-term effects were not anticipated since they have affected over time the soil structure and soil food web. Over few decades, the soil became degraded, plants are now mostly unhealthy, animals and humans experience unexplained medical conditions, and yields are going down every year. Chemicals use and tillage technology are producing the most detrimental influence on soil deterioration; as a result, their usage must be closely monitored and, if possible, avoided.

2.1 Bad practices that negatively influence the quality of agricultural soil

Farmers, working in conventional agriculture, that usually apply intensive chemical technologies, come across many harmful practices like those described below. The practices described in this section aim to draw attention to the most common activities that farmers do voluntarily or unknowingly, which may lead to soil degradation and floods.

2.1.1 Lack of efficient soil parameters evaluation

In Romania only a few farmers perform soil analysis annually, the majority of them use a standardized technology learned from books or advice from chemical companies. Soil parameters analyzed in a laboratory report do not contain enough information, the evaluation gives most often information regarding land chemistry, but ignores several important physical and biological properties. Sustainable agriculture changes the view of soil performance and soil quality [11]. Farmers need to invest more in complete soil assessment and perform some measurements by themselves, like soil acidity (pH) or soil conductivity (EC). The Haney report is another good analysis report that offers information about soil health, microbial respiration, water-extractable organic carbon, water-extractable nitrogen, etc. Haney soil test report offers farmers additional values to improve plant-available nutrients and estimate the soil health as related to carbon (C), nitrogen (N), and phosphorus (P) cycling [12].

2.1.2 Leaving the soil uncovered

Topsoil is being washed and lose its properties, microorganisms die, while its structure degrades. Uncovered soil favors natural soil erosion generated by rains, sun, or wind [13]. Soil loss of moisture and high soil temperature suppress bacteria and fungi living in the soil. In this environment, weeds germinate easily, and farmers cannot control them without using highly aggressive herbicides [14].

2.1.3 Excess of soil tillage

Tillage works as plowing and disking suppress the fungi network, appear losses the soil moisture, and may destroy soil structure [15]. Plowing is creating the hardpan at 15–30 cm deep. Hardpan is a compact layer of soil below the soil surface that inhibits roots movements through the soil [16]. Water is moving gravitationally on the hardpan, forming ponds, and soil gets salted [17].

2.1.4 Excess of synthetic fertilizers

Synthetic fertilizers suppress biology [18], contributing to soil compacting, loss of fertility, and humus total rate decrease. Plants are using only 15–30% of total inorganic fertilizers, while the rest is leaching in lakes, rivers, groundwater, etc. Accumulation of nitrogen in groundwater has different sources, being caught in irrigation lakes [19, 20]. As groundwater is the main source of drinking water, contamination poses several human health problems. At present in the United States of America, there are used 20 times more chemicals than in the American Dust Bowl period, and soil degradation continues dramatically.

2.1.5 *Excess of one specific nutrient*

Using in excess a specific nutrient especially N in a cation form, inhibits absorbing others nutrients cations as calcium (Ca), potassium (K), sodium (Na). The nutrient balance is one of the most important factors in plant nutrition [21–23], when plants receive too much N, during a 24 h photosynthesis process, N under forms of nitrate (NO_3) or ammonium (NH_4), is not transformed into proteins and became attractive for insects [24]. Excesses of N develop elongation, delay maturity, change biochemistry, cause plant stress and make plants vulnerable to drought [25].

2.1.6 *Monoculture technology*

Monoculture is not resilient to climate change, soil is losing carbon, while carbon dioxide (CO_2) is increasing in the atmosphere. Monoculture is a source of scarcity because the diversity principle is strongly affected [26]. Monoculture combined with bare soil practices decreases the fertility of agricultural lands dramatically [27].

2.1.7 *Overdraft of groundwater*

Groundwater overdraft is related to a dry aquifer, loss of water in streams and lakes, soil compacting, and polluted groundwater [27, 28].

2.1.8 *Neglecting the soil's biology*

If the biology of the soil is ignored in drought years, is a major problem, since the soil loses nutrients and water, putting plants under a high level of stress. The plant's nature enables to fill in the gap of water and nutrients. In recent decades, scientists from many laboratories have studied the interactions between microorganisms and plants, and they have concluded that the soil food web plays the most important role in plant nutrition [29].

2.1.9 *Micronutrients are underused*

In the last decades' scientific reports demonstrated that micronutrients are as important as major elements, the only difference is the needed quantity. Micronutrient deficiency is widespread in the world due to low organic matter, bicarbonate content in irrigation water, long time of drought, and imbalanced application of fertilizers. Micronutrients application contribute to plant health, soil health, and increase yield by up to 15–50% [30–32].

2.1.10 *Too much importance given to quantity and not to quality*

There is no special interest nowadays in the quality of the products obtained in conventional agriculture [33]. Healthy plants that are resistant to illnesses, insect attacks, and drought are used to produce high-quality products, while also improving yields. Highly nutritional plants have a substantial positive impact on soil health [34], animal, and human health.

3. Principles of restoring soil and plants health

Regenerative Agriculture is organic agriculture, using only natural available resources. In organic agriculture, farmers are certified if they produce non-GMO plants without using synthetic chemicals, approaching soil conservation and preservation for biodiversity. Farmers are allowed to use only inputs from certified organic agriculture. In 2018, at Rodale Institute was introduced for the first time a new higher standard for the farmers working in a regenerative system called Regenerative Organic Certified (ROC) [35]. Regenerative Agriculture is the way to sustainable farming practice, regenerate soil fertility, grow healthy plants that create healthy soils, less sensible to draught. Using methods from Regenerative Agriculture technology, carbon is sequestered in the soil, soil structure and soil fertility improve, water retention, and crop yield increase, while drought and flood ameliorate [36]. Regenerative agriculture can be defined by a holistic system approach that starts with the soil characteristics and also includes the health of the plants, animals, farmers, and community. The main aims envisage to regenerate topsoil, restoring degraded soil biodiversity, enhancing ecosystem services, improving water cycling and improving the resilience of soil to extreme weather. Regenerative Agriculture focuses on improving soil health by following four main mandatory principles and one optional. All specialists in Regenerative Agriculture accept the four principles that include soil cover, living roots, biodiversity, and minimalizing soil disturbance. The last principle which is the integration of animals is partially accepted and can be even more important in a few specific situations.

Everything plants need is cycled by soil microorganisms before becoming available to plants' roots. Earth life is based on photosynthesis, a process that transforms photonic energy into chemical energy. It varies, depending on the availability of light, water, carbon dioxide, chlorophyll concentration, and plant nutrition. Photosynthesis is the most efficient cycle and sustainable process in nature [37, 38], and it is the engine we can rely in Regenerative Agriculture. Farmers know that water, nitrogen, and high-temperature influence the photosynthesis process. During drought, plants switch from photosynthesis to photo-respiration process, when are consuming their reserve of proteins [39]. To avoid this happening, proper management has to be used that optimize nutrition. When monitoring fields frequently, one should notice nature needs [40].

3.1 Limited soil disturbance

Only a limited mechanical, chemical, and physical disturbance of soil is permitted. Tillage destroys soil structure, resulting in bare or compacted soil that is destructive to soil microorganisms and creates a hostile environment for them. Soil stability is a quantitative indicator of soil health that is based on a mix of biotic and abiotic soil parameters. The impact of physic and chemical qualities on soil resistance and resilience is mediated by the microbial community [41]. Living organisms in soil improve the structure, create pore spaces that allow water and air to infiltrate the soil. Intensive tillage destroys macro and microorganism habitat, disrupt the fungi hyphae and soil aggregate.

Synthetic fertilizers, herbicides, pesticides, and fungicides suppress life in the soil, having a negative impact on soil fertility. Inputs application disrupts the symbiotic relationships between fungi, bacteria, and plants roots. Overgrazing is a form of biological disturbance that reduces roots mass, increases soil temperature and runoff. All forms of soil disturbances affect microorganisms and diminish the soil food web.

3.2 Soil protection (defensive armor)

The principle is oriented toward keeping soil covered at all times, especially by setting up cover crops or intercropping. This is a critical step toward rebuilding soil health because bare soil is not a normal state, nature always works to cover the soil surface. When providing a natural vegetal shield, the soil is protected from wind and water erosion, while providing foods and a habitat for macro- and microorganisms [42]. It will also prevent moisture evaporation, reduces temperature, intercepts raindrops, and reduces germination of weed seeds. Soil cover offers a habitat for soil food web members that spend some of their time above ground. Keeping the soil cover on allows microorganisms to break down leftovers while recycling nutrients back into the soil.

3.3 Diversity

Nature aims for the diversity of both plant and animal species. Farmers should do the same, since monocultures are present only where humans have established them. The preservation, conservation, and restoring biodiversity should be a priority nowadays. Biodiversity is a major determinant in ecosystem stability, productivity, and nutrients dynamics. High diversity can be twice as productive as monoculture [43]. Different plant species use carbohydrates to feed certain microorganisms in return for water and nutrients via their roots. Biodiversity of plants is required to support the biodiversity of microbes. Each microorganism plays a specific role in maintaining soil health, and diversity enhances ecosystem functioning [44]. The key to improving soil health consists in a soil food web that is populated with several types of plants and animals. A fully functioning soil food web provides nutrients, water, energy, and allows the soil to express its full potential. The diversity has to be increased using crop rotation and cover crops.

3.4 Living root maintenance

Living roots have to be maintained in soil as long as possible because they are feeding soil biology by providing basic food source carbohydrates [45]. This biology feeds plants with water and nutrients, having the capacity to store nutrients and water that will be provided during drought. Farmers within conventional agriculture used to think there are 120 days to rest soil until the growing season. It is now considered wrong since living plants continue growing into early winter and break biological dormancy earlier in the spring. Their roots are feeding soil organisms and keep the biological population at a high rate. Healthy soil is dependent upon how well the food web is fed. Providing food to soil microbes helps them cycle nutrients that plants needed to grow.

3.5 Integrated animals

Nature does not function well without animal organisms. Integrated livestock into an operation provides many benefits. The major benefit is that grazing stimulates the plants to pump more carbon into the soil. This drives nutrient cycling by feeding biology, also has a major positive impact on climate change by cycling more carbon out of the atmosphere and putting it into the ground. Pasture cropping is another way of practicing regenerative agriculture for growing food and restoring degraded soil. Farmers should provide a home and habitat not only for farm animals but also for pollinators, predator insects, earthworms, and all the microbiology that drive ecosystem function.

3.6 Monitoring

Monitoring the field every day is also a key factor in keeping plants healthy. Checking the soil compaction, earthworm activity, soil structure, erosion risks, poor crop growth, etc., and keeping a recording of everyday activity helps the agricultural management system. Minimum information recorded are data, weather, fertility and irrigation program, yield, insects attack, diseases, etc.

There are different technologies according to these principles that are already used by some farmers. The most commonly used are the NO-TILL or STRIP-TILL, but they are rather used for profit maximized than for reducing drought effect and regenerating soil health. NO-TILL is studied in many countries, over a long period of years, concluded that is a big step forward [46]. However, these technologies are included in regenerative agriculture methods of growing plants during drought. A special part that is additional to these methods in regenerative agriculture, concerns breaking the hardpan and biological inoculation.

4. Methods to control draught by using the principles of regenerative agriculture

Drought stress is reduced when plants are healthy and thrive in healthy soil. For plants to overcome the draught on degraded soils, a new management strategy is required. Water, balanced nutrients, and biology are the three most important requirements for plants. Plants that are well-managed produce soil that is rich in humus. Growing healthy plants to overcome the drought and the elements that impact the process are provided in the appropriate sequence.

4.1 Weeds management

The field control has to begin in the autumn before the new agricultural year begins. Weeds like quack grass and foxtail can be found in dry clay soil, indicating calcium deficits and compact soil. Mow the grasses and compost the cuttings into the soil to help with calcium deficits. Broadleaf weeds, like ragweed, indicate copper deficiencies problem, and a phosphate/potassium imbalance. The rate between phosphate and potassium should be 2/1 for row crops and 4/1 for grass crops. Succulent weeds increase soil water capacity, replenish carbonate ions while covering the ground to protect against soil erosion. Weeds role is to deposit nutrients and metabolites in the soil or rearrange the nutrients existing in the soil. There is plenty of information in the literature about weeds role and weeds usage as a soil indicator [47–49]. This information is important to design a fertilization plan, in order to balance the nutrients. Herbicides must be avoided as much as possible since weeds get resistant to synthetic inputs, plants get unhealthy while the microbial population will decrease. Brix index in plants leaf must be measured before foliar application and 2 h after. After a few foliar applications, the crop will thrive and weeds will be attacked by insects and diseases, and not the established culture. As the nutrients are balanced, pH changes and weeds are under control.

4.2 Hardpan management

Hardpan management is the compact layer of soil just below the ground surface. Excess plowing leads to soil moisture loss by evaporation [50]. Avoiding working with moldboard plows, farmers must use instead a strip subsoil breaker in the first year to break the hardpan and apply a NO-TILL technology in the next years. Hardpan reduces the soil depth for plants roots and enhances soil waterlog. Plant roots grow in the surface layer reducing access to water and nutrients.

Well, aggregate soils are rarely found, usually, soils are crusted, compacted in layers or plow pans [51]. The agricultural year start in autumn and farmers first issue should be checking the hardpan with a penetrometer. After that, has to be measured the distance from soil surface to hardpan and hardpan thickness. If hardpan thickness is more than 5cm, then must be used a subsoil strip breaker. Soil improvement usually includes subsoil adding biological fertilization to break the hardpan and inoculate with microorganisms (bacteria, fungi) at the same time. Breaking the hardpan will allow water and nutrients to infiltrate deep in the ground, while microorganisms will keep the moisture and nutrients for a long period. Underground water and nutrients are stored naturally and through capillarity, the plants have access to water and nutrients during the drought period. In order to maintain the microbiology alive, they should be multiplied by feeding them and keeping constant moisture and temperature in the soil. In time, they will improve the soil structure, porosity, and the humus percentage will increase. In the photosynthesis process plants secret carbohydrates (sugar) and protein through the roots, which are food for bacteria and fungi. Bacteria and fungi are eaten by bigger microorganisms like nematodes and protozoa. Plants are thriving in such an environment even in drought conditions. With a restored soil food web, plants can control the water and nutrients cycling in the rhizosphere neighborhood. A restored food web reduces irrigation and tillage requirements, provides protection against pests and diseases and inhibits weeds. Pesticides and herbicides are not required, since applying these methods yields and farms profitability will be increasing.

Living life provides soil structure that resists wind and rain erosion. The first step will be in accordance with principles to use no plowing or disking, by implementing a no-till system. **Figure 2** compares three types of agricultural soil processing: in the first plan the work was performed with soil loosening equipment, in the second plan it is proposed the minimal processing technology by breaking the hardpan, and in the third plan a plowed land is highlighted.

The proposed technology within INMA institute is performed with an equipment that can be carried by an agricultural tractor, that cut the soil linearly without overturning the furrow, break the hardpan, and inoculate the ground with beneficial microorganisms. An active microorganism life restores the soil food web, which keep the pore open. This could be the first phase in rebuilding a healthy soil and ecosystem.

4.2.1 Amendment and treatments

Amendment and treatments have a significant effect on soil's physical and chemical properties and increase microbial activity. Amendments improve soil water

Figure 2.
Comparison between three types of agricultural soil processing: soil loosening equipment (first plan); minimal processing technology-hardpan breaking (second plan); plowing (third plan).

retention and soil structure as permeability, drainage, air holding capacity, etc. Soil acidity is potentially serious land degradation, acid soil is crusted and compacted, requires calcium, phosphorus, and minerals. The recommendation is to apply on soil a minimum 200 kg of lime and 200 kg of soft rock phosphates per hectare every autumn and spring during the first 2–3 years. These small quantities are recommended only in soils with degraded food web, or if microorganisms are being incorporated into the soil. Microorganisms are highly important because they break down the amendments and make them available to plants. High quantities of minerals suppress microorganisms. The amendments are spread best in autumn, before planting the cover crops and in spring before planting the main crop. Any other nutrient must be added as a result of the soil analysis. Organic amendments like compost or vermicompost have a benefic effect, increasing macro and micronutrient, organic matter, physical, and chemical soil properties like pH and EC. Humic acid found inside vermicompost, improves phytoremediation of soils contaminated with heavy metals [52]. Vermicompost soil amendments combined with foliar fertilizer, based on vermicompost, reduces the period to regenerate the soil fertility. Vermicompost can be produced in every farm, is cheap and have a tremendous effect on plants that grow during draught.

4.2.2 Mineral nutrients addition

Plants need minimum 17 mineral nutrients divided into macro- and micronutrients to grow and complete plants' life cycle. Each of the nutrients perform specific functions within the plant and the amount of each needed by the plant depends on what role the plant has each element [53, 54]. Microelements are needed in a small amount, but they are as important as macro-elements. Micronutrient deficiencies induce stress in plants, cause yield losses, resulting in poor health for animals and humans [29]. Supplying plants with micronutrients, through soil application or foliar spray, increases yields, produces higher quality, but also increases macronutrient use efficiencies. Micronutrients application is cheaper and needs less labor and transport because there are small quantities to manipulate. There are nine macronutrients nitrogen (N), phosphorus (P), potassium (K), calcium (Ca), magnesium (Mg), sulfur (S), carbon (C), oxygen (O), and hydrogen (H) from which conventional agriculture is using six, but focus only on three N, P, K. Farmers in conventional agriculture, concentrating on NPK, can deliver excellent yields in irrigated conditions or rainy years, but less quality is usually obtained. Finite products, full of nutrients, for healthy animals and humans, are obtained when plants absorb balanced nutrients. To design a fertilizer plan, one need proper knowledge of the interaction between nutrients [55]. The key to controlling the mineral nutrients is restoring the soil food web. First-year must fertilize with biofertilizers and 70% of the recommended macronutrients N and P to obtain better yields and better-quality crops, low nitrate and nitrite levels in crops [56]. In two-three years, with a soil food web restored, microorganisms will takecare of plant nutrition, bacteria will fix the nitrogen in the soil, while other specific bacteria will solubilize the needed minerals. Biofertilizer is keeping beneficial microorganisms in the soil healthy and allow plants to overcome the drought [57].

4.2.3 Organic fertilizer

Organic fertilizer is added to the soil to improve soil structure, feed both the plants and microorganisms. Microorganisms break down the organic materials and release nutrients slowly to the plants. Organic fertilizers increase soil's ability to hold water and nutrients. Solid organic fertilizer made from bat guano, fish meal, or manure can be spread on the soil before planting the main crop. Liquid fertilizer can be sprayed on soil or leaves. Chelated liquid fertilizer should be used for a slow-release technique. A cheap method is to spread the fertilizers in furrows, in this way, it will produce the same effect, but the quantity needed is much less (approximately 20–30% of the total quantity needed).

4.2.4 Seed inoculation

Seed inoculation is a cheap and beneficial tool to grow healthy plants, considering that each plant has a group of bacteria or fungi that work in association with the plant roots. The colonization of plants roots by associated bacteria and fungi result in better performance than plants colonized by the wild population of microorganisms [58]. Inoculations have to be performed for both the main crop and cover crop. Inside the cover crop, there have to be various legume seeds that can be inoculated with nitrogen-fixing bacteria. No need to fertilize the soil with nitrogen if seeds are inoculated with these types of bacteria [59, 60]. Biological control agents protect seedlings from disease as fusarium, pythium, etc. [61]. Arbuscular mycorrhizal fungi play an important role in plant growth. Corn inoculated with arbuscular mycorrhizal have a higher phosphorus absorption, increases vegetative biomass and grain yield, especially in low or medium available phosphorus [62].

4.2.5 Soil food web

Soil food web is a new model of soil fertility based on biology. This new model works better, presenting a lower cost, preventing diseases, do not pollute and use minimal chemical inputs [63]. Microorganisms are the link between water, nutrients and plants. Plants are in control of a viable soil food web, and exudates, in the form of carbohydrates and proteins, attracting specific bacteria and fungi. Bacteria and fungi consuming root exudates are at the bottom of the soil food web. Bigger microbes, nematodes, and protozoa are consuming bacteria and fungi, and are excreted as nutrients right in the rhizosphere. Protozoa and nematodes are eaten by arthropods. Arthropods may eat other arthropods or they might be eaten by snakes, birds, moles, etc. Worms, insect larvae, and moles are moving through the soil, in search of food, creating pathways and letting water and air enter. Members of the soil food web bind soil particles together, create tunnels for air and water to help create soil structure. Soil food web has a natural design and presents seven major benefits such as diseases suppression, nutrients retention, increase mineral nutrient availability to plants, improve soil structure, decomposition of toxic chemicals, production plant growth, and improve crop quality. Microorganisms and other soil food web members release root growing hormones. These growth hormones help the plant to cross the draught or a flood and increase yield.

4.2.6 Cover crops

The presence of plant cover crops in the agricultural system aids in the production of large amounts of biomass. This boosts the soil's organic content, improving fertilization. The physical, chemical and biological qualities of the soil are improved by maintaining permanent cover crops, and in time, contributing to the restoration of its health. It is recommended to use biodiversity, which include at least one species of leguminous plants. Inoculation has to be achieved with nitrogen-fixing bacteria, especially for leguminous plants. Then the amendments can be spread and may plant various cover crop seeds. Incorporation of amendments can be done with a disk harrow, while the cover crops may consist of oats, rye, buckwheat, radish, mustard, vetch, clover, etc. Plants' biodiversity will attract various bacteria and fungi, each plant species attracting its own specific microorganisms. In this way, the soil food web will be restored sooner and better. Cover crops have to be chopped or mowed in spring before full bloom, and a minimum of two weeks before planting the main crop. The cover crop will maintain the soil moisture, while soil temperature will not vary too much during drought or between day and night. After mowing, cover crops are used as mulch. Raindrop energy will be dissipated by living crops and mulch, and in this way, erosion will be under control. Cover crops are being decomposed by fungi and bacteria. Another advantage is that in winter cover crops are one of the best options to defend against topsoil loss due to erosion. If it is managed correctly, the decomposition of cover crops by bacteria and fungi provides nutrients to the main crop (cash crop), while biodiversity of cover crop suppresses weeds, prevents NO_3 leaching and produces above-ground biomass N [64, 65]. Plant diversity helps to reduce pathogens, pests, and weed invasion, reducing the need for insecticides and pesticides.

A diversified crop rotation enhances soil structure by varying the length of planting zones, allowing for better water penetration. Different crops with varying nutrient requirements, as well as waste products, will help to create a more balanced and resilient soil ecosystem. The duration of these rotations is usually between 4 and 6 years.

4.2.7 Maincrop (cash crop) management

When sowing, it is recommended to inoculate the main crop seeds with different solutions based on microorganisms and nutrients. The seeding should start in spring, two weeks after mowing the cover crops and apply foliar fertilizer during the critical point of influence. Each crop has different important phases that may be influenced by inoculation with microorganisms: when planting (to enhance germination), strengthening plant structure, growing the fruit and finishing fructification. Foliar fertilizer must contain at least calcium, manganese, boron, zinc, amino acids. Balance calcium with potassium starting filing fruit point of influence and replace calcium with potassium at the finishing fruit. Get the maximum feedback from the plant when adding biology in the fertilizer solution. A healthy plant will cross the drought. Harvest the corn seeds, but let the corn stalk on the soil to be decomposed by fungi and bacteria.

4.2.8 Foliar application

Foliar application is the most efficient and cheap way to grow healthy plants. Growing healthy plants increase the immune plants' system, get resistance to diseases and insects attack, plants can cross the drought. In order to grow healthy plants, increase the photosynthesis process from 2- to 3 times by using the right foliar fertilizer solution. Aerated compost tea is a foliar biofertilizer with a benefic impact on plant growth [66]. Inside the aerated compost tea add other nutrients needed by plants.

A complete foliar fertilizer contains clean water, mineral nutrients, microorganisms, plant bio-stimulants, bacteria bio-stimulants, fungi bio-stimulants, and inoculants. It has a synergetic effect on plant growth. Plant reaction is tremendous, especially in degraded soils.

Water is the most important ingredient in foliar fertilizer solution. Using poor-quality water can determine a loss of 50% from the effect of the foliar solution. Do not use water from ponds, lakes, or others sources without water analysis tests. Good water for foliar application has less than 70 ppm, pH between 5.2 and 6.5, electrical conductivity EC between 1.6 and 2.8 ds/mm and temperature between 58 and 78 degrees Fahrenheit. For best results use rainwater or reverse osmosis water. Municipality water is usually unsuitable for foliar recipes because of the chlorine or chloramine, with high pH and potentially hardness.

Foliar solution when humidity is high has to be applied in larger particles (not fine spray), so the liquids remain on the leaf surface a long time without drying out. Sprayers with large droplets make a huge difference. The farmer should measure the effect of the foliar solution before application and 2 h after. If Brix reading is 2 points higher, the foliar solution is good and could apply on the entire field. A diagram of the Brix index reading should be done for every crop. Around 8 o'clock in the morning, after collecting healthy old mature leaf samples from 10 to 20 plants, they are squeezed on a refractometer.

In the Brix diagram example, the values are starting from 5, increasing to 9 at the first foliar application, but dropping after a few days to 6. After repeating the foliar application, the Brix index increases to 11, but drop in a few more days to 7. Every time when a good foliar application is applied, the Brix readings are higher and has been found that when Brix values are over 12, the plants present a health status that helps them overcome more easily the drought.

4.2.9 Sap analysis

A refractometer gives general information about plant health, but for more information including nutrients balance, a sap analysis is necessary. Plant sap analysis provides 21 nutrients parameters values that enable farmers to optimize the crops' fertilization plan. The information is valuable because the uptake of plant nutrients are revealed in a few hours, the increasing performance can be tracked graphically, similar to the example shown in **Figure 3**. To a better understanding, one can compare a sap plant analysis with human blood analysis. A plant sap analyze tells the current

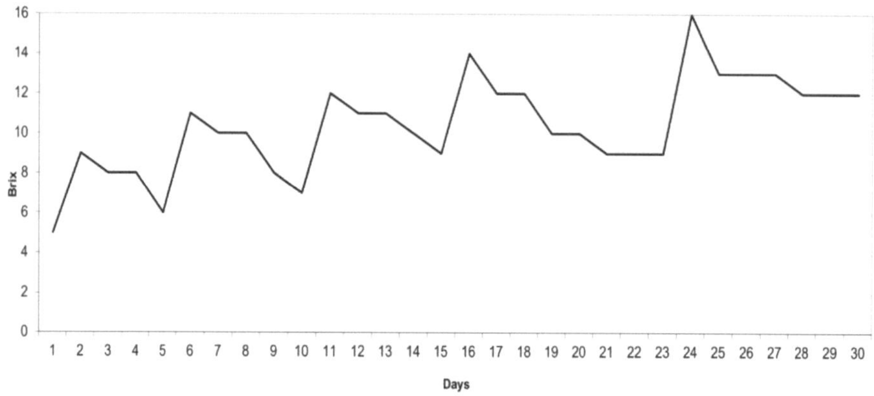

Figure 3.
Example of the variation of the Brix index for tomato juice, for a period of 30 days.

uptake of nutrients, excesses or deficiencies of nutrients long term before can be seen on a plant leaf, plant reserves of nutrients, nutrient imbalance in soil, what nutrient plant can use at that moment for its own growth, or even fruit quality [67]. Sap analysis laboratory in less than one week will provide the analysis sheet with a fertilizer plan recommended. A balanced mineral uptake increases plants' health that gets resistant to diseases and insects attack and crosses the drought.

4.2.10 Water management

Water is the most important nutrient for plants. A source of water is critical for drought years, but as long as the regenerative methods presented are met, plants can cross the draught without an irrigation water source. Knowledge of irrigation water quality is critical in understanding long-term soil management. The most influential water parameter is the salinity measured by electrical conductivity EC [68]. High sodium related to calcium and magnesium contends, in irrigation water, causes surface crusting, pore plugging, swelling, and dispersion of soil clay. The acidity or basicity of irrigation water is expressed as pH. Normal irrigation pH is between 6.5 and 8.5. Specific ions like boron, sulfate, chloride, and nitrate, may affect plants grown. An irrigation water analysis is required.

4.2.11 Crop rotation

Keep a crop rotation, with cover crop intermediate, for minimum 3 years after starting your regeneration soil program. After concluding that the soil food web is active and the soil is healthy crop rotation is not as important anymore, since biology will take care of plant nutrition and will suppress diseases, insects attack will decrease.

4.2.12 Aerated compost tea

Aerated compost tea, produced by a compost tea brewer, allow microorganisms to be extracted from compost and multiplied. The result consists in beneficial aerobic microorganism production that provides plants with nutrients and helps build the soil food web [69]. The tea is used for spraying both the leaf and soil.

Vermicompost is also used to avoid pathogens. Red worm castings are free of pathogens. There are farmers, involved in regenerative agriculture, buying different products that contained few families of fungi and bacteria, but inside aerated compost tea there are thousands of families. A compost tea brewer can be purchased or can easily be built. All a farmer needs is a tank, an air pump, a hose, and an air splitter distributor. To brew the compost needs clean water, vermicompost, mineral nutrients and bio-stimulants for plants, bacteria and fungi. Brew all these ingredients for 24–36 h, then measure the pH and EC. If pH is higher than 6.5 must add 100–300 ml of vinegar and measure again. When measuring EC a few hours before stopping the air pump, If the values are too low must add more vermicompost. The tea has to be used within 4 h after the air pump stops, to improve the synergetic effect on plants and soil [70].

4.2.13 Managing livestock

Good and efficient management of animal grazing can rebuild soil health. This is a way for a healthy ecosystem, farm profitability, human health, food system resilience. Studies that use a complementary approach to animal husbandry with organic farming use found that adopting some grazing strategies could regenerate the soil and make them more profitable. Holistic management of livestock management includes grazing, land, and financial planning and ecological monitoring.

4.2.14 Agroforestry

Agroforestry can provide suitable tools for landscape restauration because it can enhance physical, chemical, and biological soil characteristics. Agroforestry is restoring and increasing land productivity because the presence of the trees can fix nitrogen, stabilize the soil, reduce soil erosion, increase fertility, and regulate water available in degraded lands.

Trees increase fertility by retrieving nutrients from deeper soils and adding them to the soil surface through the leaf litter. Because of their deep root system, trees prevent nutrients from leaching, combat soil salinization, and acidification. The use of trees with fixing-nitrogen bacteria is increasing crop productivity. Experiments in Zambia, for example, showed that maize yields increased by 88–190% when grown in an agroforestry system under a canopy of *Faidherbia Albida Trees* (FAO report Agroforestry for landscape restauration).

Trees can reduce and prevent soil erosion planted in windbreaks trees protect soil from erosion and increase yield.

Agroforestry buffer strips increase water runoff, and soil evaporation and increase water infiltration and water retention capacity, helping plants to cross the drought.

4.2.15 Instrumentation

Minimum instrumentation required to grow healthy plants and cross the draught more easily is the penetrometer, refractometer, pH-meter, EC-meter. A penetrometer is the first instrument to be used in an agricultural season to measure soil compaction. The penetrometer is a device used to measure the resistance of soil to a vertical force. The penetrometer can determine the depth of the hardpan and help producers to determine if a subsoil is in need.

Refractometer measure Brix index values for liquids. Brix values indicate the total soluble solids. The refractometer is widely used in measuring the quality of the grapefruits and the time to harvest. The refractometer can be used to evaluate the overall assessment of plant health. Healthy plants with a minimum 12 Brix readings are resistant to diseases and insect attacks.

Soil pH-meter is used to measure the acidity or alkalinity of soil. The values give information about the balance of the nutrients found in the soil. However, can be also used as a pH meter for liquids, and determine pH when adding 5 parts distilled water on 1 part of the soil.

EC (electrical conductivity) meter is used especially to estimate salinity levels. A high level of salinity reduces the plant's ability to take up water. For assessment is being used 5 parts distilled water and 1 part soil to determine the values of salinity in mSiemens/cm. In clay soils, values are between 0.2 and 1.0 mS/cm, but different plants tolerate different values.

4.2.16 Drought

During drought, when air temperature became too warm, plants switch from photosynthesis to photo-respiration and begin consuming their inside proteins. Healthy plants with a waxy sheen, on the leaf surface, have a cooler leaf temperature than plants with a lack nutritional integrity. Foliar applications with teas made from compost, with the addition of 3 L of molasses per hectare, during and after the drought, is a very good practice.

5. Conclusion

Regenerative agriculture is focused on farming techniques with the primary goal of regenerating the land, particularly increasing the organic composition in order to improve fertility. This strategy conserves and restores soil organic matter, thus, influencing the development and prosperity of micro- and macro-organisms with beneficial results against soil erosion and drought.

Farmers may be forced to adopt unsustainable practices due to economic pressures, as they rarely have enough ability to deal with the conditions imposed by larger corporations, that control prices and credit. As a result, agricultural policies must be implemented at the national level to assist farmers and ensure they are not compelled to deplete the resource that provides them with a means of subsistence.

Regenerative agriculture is based on a holistic approach that places the land at the core of the process to produce efficiently and sustainably a synergy between the soil, the animal world, and the plant world. This enables the development of food chains between all three ecosystems, while the restoration of soil health is ensured by the balance and diversity of species found within the environment.

Climate change is no longer a myth, but a fact and the consequences are becoming increasingly severe every day, influencing the drought phenomena. Every year, topsoil is leaching, soil gets compacted, crusted, loses the ability to supply nutrients and water to plants. Degraded soils, in drought conditions, are not able to support plants with the required water and nutrients, while yields decrease dramatically. In order to reduce the drought effect, farmers have to integrate their use of regenerative agriculture principles and methods, focusing on growing healthy plants and getting rewarded with good yields and increased farm profitability.

Water retention in agricultural lands is associated with soil organic carbon and is influenced by soil health. Soil organic carbon increases the percentage of water retention because carbon acts like a sponge that absorbs moisture. Regenerative management practices such as minimum tillage, cover crops, inoculation with microorganisms, mulching practices, nutrients cycling, maintenance of an optimal balance of organic fertilizers, foliar application, and other methods help to increase soil organic carbon. This strategy restores degraded soils, enhances biomass production, purifies groundwater, reduces the rate of CO_2 emissions into the atmosphere and increases the percentage of water being retained in the soil.

An active soil food web is the link between water, nutrients, and plants. Healthy soils have an active soil food web that presents many benefits such as diseases suppression, nutrient retention, improve soil structure, making mineral nutrients available to plants, decomposition of toxic materials, improve crop quality. Soil food web works in synergy with plants and helps crops to overcome more easily drought or floods.

The primary goal of this technology is to grow healthy plants on a worldwide scale. Healthy plants achieve synergies with the soil and improve its health, recover carbon in the soil, increase water retention, and improve soil structure and nutritional status. Drought years will be more profitable for farmers using regenerative agriculture technology, since organically grown cereal prices will be higher, resulting in greater average yields. In a short period of time, farmers using regenerative agriculture technology will spend less money, yields will grow, profitability will increase, soils will regenerate, and drought years will become less risky.

Author details

Eugen Popescu, Florin Nenciu* and Valentin Nicolae Vladut
INMA Bucharest—National Institute for Research-Development of Machines and Installations Designed for Agriculture and Food Industry, Bucharest, Romania

*Address all correspondence to: florin.nenciu@inma.ro

References

[1] Gomiero T. Soil degradation, land scarcity and food security: Reviewing a complex challenge. Sustainability. 2016;**8**:281. DOI: 10.3390/su8030281

[2] Crawford E. 3 basic tenets of regenerative farming could improve soil health, food production and bottom line. Food navigator-usa.com. 2018. Available from: https://www.foodnavigator-usa.com/Article/2018/08/17/3-basic-tenets-of-regenerative-farming-could-improve-soil-health-food-production-and-bottom-lines

[3] Coursework Stanford University. Topsoil Erosion 2015. un report 2015 humanity is losing 0.3% of food per year. Available from: http://large.stanford.edu/courses/2015/ph240/verso2/

[4] Hornbeck R. The enduring impact of the American Dust Bowl: Short- and long-run adjustments to environmental catastrophe. American Economic Review. 2012;**102**:1477-1507

[5] Gretton P, Salma U. Land degradation and the Australian agricultural industry, Staff information paper, Industry Commission. GPO Box 84, Canberra ACT 2601: Commonwealth of Australia, Australian Government Publishing Service, AGPS; 1996. Available from: https://www.pc.gov.au/research/supporting/land-degradation/landdegr.pdf

[6] Lal R, Hall GF, Miller FP. Soil degradation: I. Basic processes. LDD Land Degradation & Development. 1989;**1**: 51-69. DOI: 10.1002/ldr.3400010106

[7] Oldeman RL. Global Extend of soil Degradation. Bi-Annual Report. ISRIC Bi-Annual Report. 1992. pp. 19-36. Available from: https://library.wur.nl/WebQuery/wurpubs/fulltext/299739

[8] Butzer KW. Environmental history in the Mediterranean world: Cross-disciplinary investigation of cause-and-effect for degradation and soil erosion. Journal of Arheological Science. 2005;**32**:1773-1800. DOI: 10.1016/j.jas.2005.06.001

[9] McGarry D. Tillage and Soil Compaction. Conservation Agriculture2003. pp. 307-316. Available from: https://link.springer.com/chapter/10.1007/978-94-017-1143-2_37

[10] Savci S. Investigation of effect of chemical fertilizers on environment. APCBEE Procedia. 2012;**1**:287-292. DOI: 10.1016/j.apcbee.2012.03.047

[11] Romig DE, Garlynd MJ, Garris MF, Mc Sweeney K. How farmers assess soil health and quality. Journal of Soil and Water Conservation. 2012;**1**:287-292. DOI: 10.1016/j.apcbee.2012.03.047

[12] Haney RL, Haney EB, Smith R, et al. The soil health tool—Theory and initial broad-scale. Applied Soil Ecology. 2018;**125**:162-168

[13] Shiva V. Monocultures of the Mind: Perspectives on Biodiversity and Biotechnology. Available from: https://books.google.ro/books?hl=en&lr=&id=QcstWYIcbHkC&oi=fnd&pg=PA5&dq=shiva+monocultures+of+the+mind&ots=vBJ6J4d6sM&sig=Z

[14] Habib U, Eroglu G, et al. Using high resolution images and elevation data in classifying erosion risks of bare soil areas in the Hatila Valley Natural Protected Area, Turkey. Stochastic Environmental Research and Risk Assessment. 2010;**24**:699-704 Available from: https://link.springer.com/article/10.1007/s00477-009-0356-5

[15] Teasdale JR, Daughtry CST. Weed Suppression by Live and Desiccated Hairy Vetch (*Vicia villosa*). Cambridge: Cambridge University Press; 2017. Available from: https://www.cambridge.org/core/journals/weed-science/article/abs/weed-suppression-by-live-and-desiccated-hairy-vetch-vicia-villosa/91

[16] Baumhardt L. Dust Bowl Era. Encyclopedia of Water Science 2003. Available from: https://www.baymeadowscharter.org/ourpages/auto/2019/9/16/54520179/Lithosphere%20KQ4_%20Dust%20Bowl%20Case%20Study.pdf

[17] Chen Y, Cavers C, Tessier S, Monero F, Lobb D. Short-term tillage effects on soil cone index and plant development in a poorly drained, heavy clay soil. Soil and Tillage Research. 2004;**82**:161-171

[18] Tsubo M, Fukai S, Basnayake J. Effects of soil clay content on water balance and productivity in Rainfed Lowland Rice Ecosystem in Northeast Thailand. Plant Production Science. 2007;**2007**:10. Available from: https://www.tandfonline.com/doi/abs/10.1626/pps.10.2

[19] Tripath S, Srivasta P, Devi RS. Chapter 2—Influence of synthetic fertilizers and pesticides on soil health and soil microbiology. In: Agrochemicals Detection, Treatment and Remediation. 2020. pp. 25-54. DOI: 10.1016/B978-0-08-103017-2.00002-7

[20] Power JF, Scheper S. Nitrate contamination of groundwater in North America Agriculture. Ecosystems & Environment. 1989;**26**:165-187

[21] Andensen A. Science in Agriculture. ISBN-13: 978-0911311358. Available from: https://www.amazon.com/Science-Agriculture-Advanced-Methods-Sustainable/dp/0911311351

[22] Fageria VD. Nutrient interactions in crop plants. Journal of Plant Nutrition. 2006;**24**:1269-1290. DOI: 10.1081/PLN-100106981

[23] Novoa R, Loomis RS. Nitrogen and plant production. Plant and Soil. 1981;**58**:177-204. Available from: https://link.springer.com/article/10.1007/BF02180053

[24] Goyal SS, Huffaker RC. Chapter 6. Field lab earth podcast. In: Nitrogen Toxicity. 2001. DOI: 10.2134/1990.nitrogenincropproduction.c6

[25] Shiva V. Monocultures of the Mind. Perspectives on Biodiversity and Biotechnology. 1993. Available from: https://books.google.ro/books?hl=en&lr=&id=QcstWYIcbHkC&oi=fnd&pg=PA5&dq=shiva+monocultures+of+the+mind&ots=vBJ6K1l6sK&sig=o4EjfW2z7L3oW2zxP9xMP0YTgMw&redir_esc=y#v=onepage&q=shiva%20monocultures%20of%20the%20mind&f=false

[26] Ewel JJ, Mazzarino MJ, Berish CW. Tropical soil fertility changes under monocultures and successional communities of different structure. Ecological Aplication. 1991;**11991**:289-302. DOI: 10.2307/1941758

[27] Zektser S, Loáiciga HA, Wolf JT. Environmental impacts of groundwater overdraft: Selected case studies in the southwestern United States. Environmental Geology. 2004;**47**:396-404. Available from: https://link.springer.com/article/10.1007/s00254-004-1164-3

[28] Collins W, Qualset CO. Biodiversity in Agroecosystems. West Palm Beach, FL: CRL Press LLC; 1999

[29] Malakouti MJ. The effect of micronutrients in ensuring efficient use of macronutrients. Turkish Journal of

Agriculture and Forestry. 2008;**32**:215-220. Available from: https://journals.tubitak.gov.tr/agriculture/abstract.htm?id=9548

[30] Römheld V. Marschner H. Function of Micronutrients in Plants FIELD, LAB, EARTH PODCAST. 1991. Chapter 9. doi:10.2136/sssabookser4.2ed.c9

[31] Allowey BJ. Micronutrient Deficiecies in Global Crop Production. United Kingdom: Springer; 2008. pp. 1-39 https://books.google.ro/books?hl=en&lr=&id=_55yK0hj67IC&oi=fnd&pg=PR2&dq=brian+micronutrients+deficiencies&ots=WuW6zive7_&sig

[32] Douglas E, Romig M, Garlynd J, Harris RF, McSweeney K. How farmers assess soil health and quality. Journal of Soil and Water Conservation. 1995;**50**:229-236

[33] Rodale Institute blog. Available from: https://rodaleinstitute.org/why-organic/organic-basics/regenerative-organic-agriculture/

[34] Rhodes CJ. The imperative for regenerative agriculture. Research Progress. 2017. DOI: 10.3184/003685017X14876775256165

[35] Dougherty B. Regenerative Agriculture: The Path to Healing Agroecosystems and Feeding the World in the 21st Century Nuffield International. 2019. Available from: https://www.nuffieldscholar.org/sites/default/files/reports/2018_US_Brian-Dougherty_Regenerative-Agriculture-The-Path-To-Healing-Agroecos

[36] Galston AW. Photosynthesis as a basis for life support on earth and in space. BioScience. 1992;**42**:490-493. Available from: https://www.jstor.org/stable/1311878

[37] Jensen PJD, Lambreva MD, Plumere N. Photosynthesis at the forefront of a sustainable life. Crop Biology and Sustenability. 2014;**2**. DOI: 10.3389/fchem.2014.00036

[38] Ashraf M. Inducing drought tolerance in plants. Biotechnology Advances. 2010;**28**:169-183

[39] Brown G. Dirt to Soil. Vol. 7. ISBN13 (EAN): 9781603587631

[40] Nunes MR, Karlen DL, Moorman TB. Tillage intensity effects on soil structure indicators—A US meta-analysis. US Sustainability. 2020;**12**. DOI: 10.3390/su12052071

[41] Griffiths BS, Philippot L. Insights into the resistance and resilience of the soil microbial community. FEMS Microbiology Reviews. 2013;**37**:112-129

[42] Cammeraat LHA, Cimeson A. Deriving indicators of soil degradation from soil aggregation studies in southeastern Spain and southern France. Geomorphology. 1998;**23**:307-321

[43] Tilman D, Isbell F, Cowles JM. Biodiversity and ecosystem functioning. Annual Review of Ecology, Evolution, and Systematics. 2014;**45**:471-493. DOI: 10.1146/annurev-ecolsys-120213-091917

[44] Cheng W, Coleman DC. Effect of living roots on soil organic matter decomposition. Soil Biology and Bichemistry. 1990;**22**:781-787. DOI: 10.1016/0038-0717(90)90157-U

[45] Francis CA, Harwood RR, Parr JF. The Potential for Regenerative Agriculture in the Developing World. Cambridge University Press; 2009. Available from: https://www.cambridge.org/core/journals/american-journal-of-alternative-agriculture/article/abs/potential-for-regenerative-agriculture

[46] Franzluebbers AJ, Hons FM, Zuberer DA. Tillage and crop effects on seasonal dynamics of soil CO_2 evolution, water content, temperature, and bulk density. Applied Soil Ecology. 1995;**2**:95-109

[47] Andersen A. Science in Agriculture. 2nd ed. Vol. 12. Greeley, CO, USA: ACRES USA; pp. 189-203

[48] Walters C. Weeds Control without Poison. Greeley, Colorado, USA: ACRES USA; 1999

[49] Anderson A. The Anatomy of Life & Energy. Vol. 4. Greeley, CO, USA: ACRES USA; p. 51

[50] Gerard CJ, Bloodworth ME, Burleson CA, Cowley WR. Hardpan formation as affected by soil moisture loss. Soil Science Society Journal. 1961;**25**:460-463. DOI: 10.2136/sssaj1961.03615995002500060013x

[51] Chen Y, Cavers C, Tessier S. Short-term tillage effects on soil cone index and plant development in a poorly drained, heavy clay soil. Soil and Tillage Research. 2005;**82**(2):161-171. DOI: 10.1016/j.still.2004.06.006

[52] Anghelova VR, Akovai VI, Artinova NS, Ivanov KI. The effect of organic amendments on soil. Bulgarian Journal of Agricultural Science. 2013:958-971

[53] Barker AV, Pilbeam DJ. Handbook of Plant Nutrition. Boca Raton, Florida, USA: CRC Press; 2015

[54] Jones C. Jacobsen J. Plant Nutrition and Soil Fertility. Nutrient Management Module nr.2. Available from: https://citeseerx.ist.psu.edu/viewdoc/download?doi=10.1.1.576.5603&rep=rep1&type=pdf

[55] René P, Rietra J, Heinen M. Effects of nutrient antagonism and synergism on yield and fertilizer use efficiency communications in soil science and plant analysis. 2017;**48**:1895-1920. DOI: 10.1080/00103624.2017.1407429

[56] El-Yazied AA, Selim SM. Effect of reducing n, p mineral fertilization levels combined with biofertilizer on growth, yield and tuber quality of potato plants. Journal of Plant Production. 2007;**32**:2957-2982

[57] Zainuddin N, Keni MF, Ibrahim SAS, Masri MMM. Effect of integrated biofertilizers with chemical fertilizers on the oil palm growth and soil microbial diversity. Biocatalysis and Agricultural Biotechnology. 2022;**39**. DOI: 10.1016/j.bcab.2021.102237

[58] Hartman G, Khadka R, Uphoff N. Benefits to plant health and productivity from enhancing plant microbial symbionts the plant holobiont Volume II: Impacts of the rhizosphere on plant health. Frontiers in Plant Science. 2021;**11**:1-21. DOI: 10.3389/fpls.2020.610065

[59] Mao W, Lewis JA, Hebbar PK, Lumsden RD. Seed Treatment with a Fungal or a Bacterial Antagonist for Reducing Corn Damping-off Caused by Species of *Pythium* and *Fusarium*. APS Publications. 2007;**81**(5):450-454. DOI: 10.1094/PDIS.1997.81.5.450

[60] Bressan W. Biological control of maize seed pathogenic fungi by use of actinomycetes. Bio-Control. 2003;**48**:233-240. Available from: https://link.springer.com/article/10.1023/A:1022673226324

[61] Bennett MA. The Use of Biologicals to Enhance Vegetable Seed Quality: Seed Technology. Vol. 20. Association of Official Seed Analysts; 1998. pp. 198-208. Available from: https://www.jstor.org/stable/23433023

[62] Gomez SC, Fonseca CR, Giachini SE. Yield increase of corn inoculated with a commercial arbuscular mycorrhizal inoculant in Brazil. Santa Maria RS Brazil, Centro de Ciências Rurais:Crop Production. 2020. DOI: 10.1590/0103-8478cr20200109

[63] Lowenfels J, Lewis W. Teaming with microbes. Portland, Oregon: Timber Press; 2010

[64] Finney DM, White CM, Kaye JP. Biomass production and carbon/nitrogen ratio influence ecosystem services from cover crop mixtures. Agronomy Journal. 2016. DOI: 10.2134/agronj15.0182

[65] Mitchall JP, Shrestha A, Mathesius K. Cover cropping and no-tillage improve soil health in an arid irrigated cropping system in California's San Joaquin Valley, USA. Soil and Tillage Research. 2017;**165**:325-335. DOI: 10.1016/j.still.2016.09.001

[66] Otero M, Kepa IS, Gonzales-Murua C. Dunabeitia quality assessment of Pinus radiata production under sustainable nursery management based on compost tea. Journal of Plant Nutrition and Soil Science. 2019. DOI: 10.1002/jpln.201800309

[67] Esteves E, Locatelli G. Sap analysis: A powerful tool for monitoring plant nutrition. Horticulturae. 2021;**7**(11):426. DOI: 10.3390/horticulturae7110426

[68] Bauder TA, Waskom RM, Sutherland PL, Davis JG. Irrigation Water Quality Criteria. Colorado State University, U.S., Department of Agriculture and Colorado counties cooperating; 2011

[69] Ferguson J. The role of organic materials in gardening. Lone Star Regional Native Plant Conference 2003. Chireno. Available from: https://d1wqtxts1xzle7.cloudfront.net/68954321/viewcontent.pdf?1630356850=&response-content-disposition=inline%253B+filename%253D2nd_Lo

[70] Ingham ER. The compost tea brewing manual. Soil Food International.; 2022. Available from: https://www.soilfoodweb.com.au/shop/manuals/manuals/compost-tea-brewing-manual-5th-edition-detail

Section 3

Drought and Plants

Chapter 5

Drought Stress: Manifestation and Mechanisms of Alleviation in Plants

Kousik Atta, Aditya Pratap Singh, Saju Adhikary, Subhasis Mondal and Sujaya Dewanjee

Abstract

Drought can be referred to as a meteorological period without significant rainfall and it is one of such major abiotic stresses that contributes to a huge reduction in crop yield throughout the world. Plant shows a broad range of physiological, morphological, and biochemical changes such as reduced photosynthetic accumulation, altered gene expression, etc. Under the drought stress which ultimately causes reduced growth as well as poor grain yield. Drought stressconditions trigger production of ROS, which disrupts the dynamic balance between ROS production and ROS scavenging systems and its accumulation depends on the intensity as well as duration of water stress, and it varies among species. A plant species that has a better inherited genetic response allowing it to rapidly deploy its antioxidant enzymatic and non-enzymatic defense system, can tolerate drought better than a plant species with a poor antioxidant defense system. Furthermore, enzyme and protein encoding drought specific genes have the ability to enhance drought tolerance. These two enzymatic and genetic engineering strategies are unique and vital tools, which can be used to help alleviate the world's future problems related to energy, food, and environmental stresses, particularly drought. This chapter attempts to discuss developments in understanding effects of drought stress and underlying mechanisms in plants for its alleviation.

Keywords: ABA signaling, antioxidant, drought, ROS, stress

1. Introduction

Any inimical condition or substance that affects plant's metabolism, growth and development is referred as stress. Basically, stress is an altered physiological condition caused by different living and non-living factors which disturb the equilibrium. Plants are frequently posed with a plethora of stress conditions such as drought, salinity, heat stress, low temperature, heavy metal toxicity, flooding and extremes of soil pH. Plants also face challenges from biotic factors like pathogens, insects etc. These types of abiotic and biotic factors limit plants growth and productivity. The non-living variable must impact the environment beyond its normal range of variation to unfavorably affect the population performance or individual physiology of the organism in a significant way.

Drought is a meteorological term and defined as a period without significant rainfall. Generally, drought stress occurs when the available soil-water becomes scanty and atmospheric conditions cause continuous loss of water by transpiration or evaporation. Water deficit is one of the major abiotic stresses, which adversely affects crop growth and yield. These changes are mainly associated with altered metabolic functions, one of those is either loss of or diminished synthesis of photosynthetic pigments, uptake and translocation of ion, carbohydrate biosynthesis, nutrient metabolism and synthesis of growth promoters. These changes in the metabolic functions and synthesis of photosynthetic pigments are closely related to biomass production in plant [1]. A common adverse effect of water stress on crop plants is the reduction in fresh and dry biomass [2]. Plant productivity under moisture stress is strongly associated with the processes of dry matter partitioning and temporal biomass distribution [3]. Previous study about different crop species faces huge yield reduction due to drought stress (**Table 1**). We have aimed to discuss the crops' response and adaptive mechanisms to combat drought stress and also genetic interventions which may help developing cultivars suitable for water-scarce conditions.

2. Physiological changes during drought stress

During drought, Water scarcity occurs generally because of absence of water in the soil. But Physiological drought caused both lack of water in the soil, and also occurs when excess water is present in the soil. Thus, physiological drought is a situation where the plant cannot receive water [15, 16]. The responses of plants to water stress are diverse and may involve the contribution of various defense mechanisms or modification of physiology, morphology, anatomy, biochemistry, as well as short and long-term developmental and growth related adaptation processes [17].

Crop	Yield reduction (%)	References
Rice	53–92	[4]
Maize	79–81	[5]
Barley	49–57	[6]
Chickpea	45–69	[7]
Pigeonpea	40–55	[8]
Soybean	46–71	[9]
Sunflower	60	[10]
Potato	13	[11]
Canola	30	[12]
Cowpea	55–65	[13]
Wheat	64.46	[14]

Table 1.
Yield reduction owing to drought stress in different crops.

Physiological reactions to moisture stress provides some escape mechanisms to the water stress comprise physiological and morphological adaptations [18]. Decreased leaf area (**Figure 1**), reduced stomatal number and conductance, enlargement of root system, increased leaf thickness, and leaf folding to lessen evapotranspiration are strictly associated with an adaptive response [17, 19–21]. Plant growth and productivity decreased under moisture stress, which are caused by alterations in plantwater relations, CO_2 assimilation reduction, membrane damage of affected tissues, cellular oxidative stress, and inhibition of enzymes activity.

Plants can alter water relations to continue cellular mechanisms under drought stress conditions. Plants show osmotic adjustment by accumulating and integrating compatible solutes likely, proline, sugars and free amino acids [22]. Maintenance of turgor pressure as well as cell volume at low water potential is facilitated by osmotic adjustment and is vital for metabolic functions. Osmotic adjustment also plays role in recovery of metabolic activities post drought stress [23]. Previously, there are lot of studies investigated which showed the recovery of photosynthesis from moisture stress in various crop species and also recovered from drought stress in terms of oxidative stress, membrane stability index and antioxidative mechanisms [16, 24]. Osmolytes also have a significant role in drought stress recovery.

Drought stress at higher intensity decreases the activities of photosynthetic enzymes as well as leaf chlorophyll content which ultimately hampers the process of photosynthesis [20, 25]. Chlorophyll *a*/*b* proportion and synthesis of leaf chlorophyll altered during drought stress. A lower content of chlorophyll (**Figure 2**), inactivation of key proteins linked to the photosynthesis process, and alteration of thylakoid membranes happen as a result of drought stress. The decline in chlorophyll content is due to over production of O_2^- and H_2O_2 production, which ultimately results significant chlorophyll degradation and lipid peroxidation. During drought stress, in stomata and mesophyll cell the CO_2 conductance declined as the decrease in the photosynthetic process. The decrease in photosynthetic activity also may be because

Figure 1.
Effects of different levels of drought stress on ricebean seedlings.

Figure 2.
Visual effects of drought stress in rice. Source: [26].

of the reduction of stomatal movement [27]. Rubisco activity greatly affected by the loss of CO_2 uptake, similarly it decreases the activity of sucrose phosphate synthase, nitrate reductase and RuBP production [20, 28]. Decline in photosynthetic activity, loss of photosystem II photochemical efficiency, reduction in chlorophyll content and alteration in stomatal movement results the reduction in plant productivity. As a consequence of reduction in photosynthetic activity in drought stress, it dismantles the production of carbohydrate in various way likely prevents the transport of sucrose into sink organs and reduces the level of sucrose in leaves, which in turn limits reproductive development. The free sugars and different metabolites thereof support plant growth under drought, and take up osmolytic role and compatible solutes to mitigate the drastic effect of the stress [29, 30].

The relative leaf water content (RLWC) is an estimate of leaf's hydration status relative to its maximal water holding capacity at full turgid state. The relative leaf water content (RLWC) is one of the reliable parameters to know the water status in plants and it decreases gradually with increases in the severity of drought stress conditions. The decline of RLWC as a response to osmotic stress was earlier reported by several investigators under different stress conditions [31–34]. The physiological traits considered for evaluating drought stress tolerance include root trait characteristics (root length, root density, root biomass, root length density, delayed canopy wilting (DCW) and leaf pubescence density (LPD) [35], delayed leaf senescence (DLS) [36], and recovery ability after wilting (RAW) [37]. Drought stress drastically affects seed germination and decreases the speed of germination (**Figure 3**). Apart from these, stomatal conductance, chlorophyll content and use of carbon isotope discrimination are also effective screening methods for drought stress tolerance and has been used for some food legumes.

Figure 3.
Ricebean response to varying levels of PEG as drought induction agent.

3. Plants adaptive responses to drought stress

Plants have developed various adaptive mechanisms conferring tolerance to drought stress induced adversities through evolution [38]. Their survival strategies for drought stress can broadly be classified as escape, avoidance and tolerance. Hence, their drought stress response varies from molecular to plant level [39]. The mechanisms of plant escape, avoidance and tolerance (**Figure 4**) against drought stress are discussed as follows.

3.1 Escape, avoidance and tolerance mechanisms

To escape the pernicious effects of drought stress on plant health and productivity, some plants utilize mechanisms involving shortening of the life cycle by rapid plant development, self-reproduction, and seasonal growth before the beginning of the drought season (**Figure 4**) [40]. Among all, early flowering is perhaps the best possible escape adaptive mechanism in plants [41]. However, this mechanism can connote a considerable reduction in the plant's growing period compromising plant productivity in some cases [42].

In avoidance strategy, high plant water potential is maintained through transpiration loss reduction and the increased water uptake from well-established root systems [43]. Xeromorphic features such as the presence of hairy structure on leaves and cuticles in some cases do help to maintain high water potentials in plant tissues [44]. It is notable that overdevelopment of these structures may lead to reduced productivity and reduced decreased size of vegetative and reproductive parts [45]. On the contrary, an adaptive tolerance mechanism at the photosynthetic level involves reductions in the plant's total leaf area and limited expansion of new leaves. Likewise,

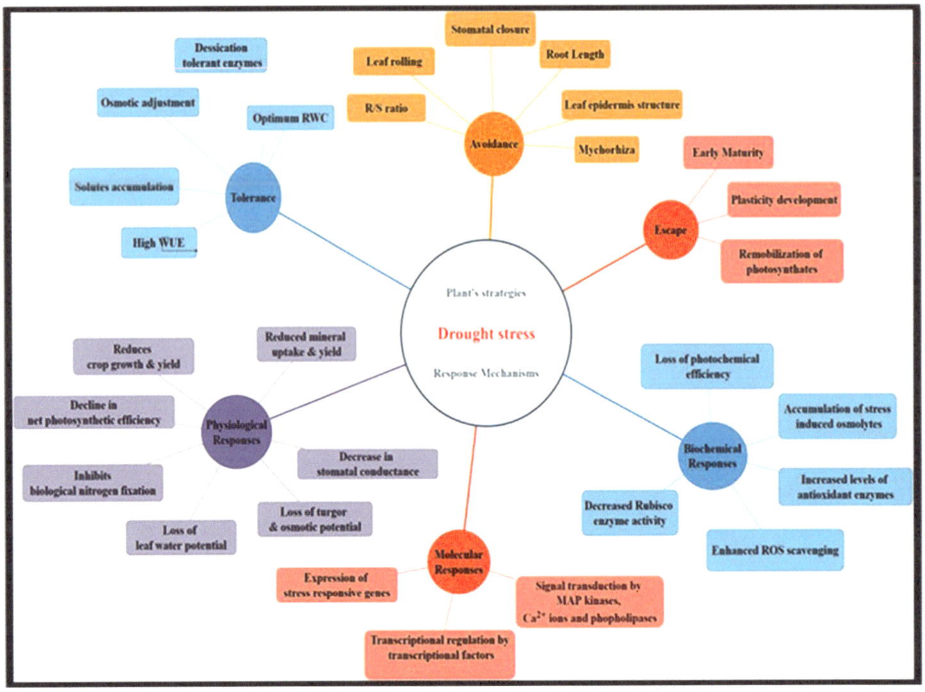

Figure 4.
Overall drought stress response in crop species.

trichrome production on leaves is an attribute that enables the plant to tolerate water deficits in dry environments [46]. There is an increase in rate of light reflection in the leaf reducing the leaf temperature as well as trichomes provide additional layer of resistance to the water loss thereby reducing the rate of water loss through transpiration [47]. Changes in root system-size, density, length, proliferation, expansion and growth rate, constitute the preliminary strategy for drought-tolerant plants to cope against drought [48]. Osmotic adjustment, antioxidant defense mechanism, metabolic and biochemical dynamics of stomatal closure, solute accumulation and increment in root shoot ratio are other common strategies that aid to drought stress resilience [49].

4. Biochemical responses to drought

4.1 Oxidative damage

Drought stress triggers an array of biochemical mechanisms including fluidity of the plasma membranes, osmolytes production, lipid peroxidation, reactive oxygen species (ROS) generation, rigidity of the cellular membranes and activation of different enzymes which are involved in oxidative defense system [50, 51]. Previously, in various crop species ROS generation instigates significant damage to cellular components and also causing damages to lipid peroxidation, proteins [52]. The drought stress induced ROS generation had calamitous effects on lipid membrane and protein. Among all the ROS superoxide radical ($O2^{\bullet -}$), hydrogen peroxide (H_2O_2), singlet

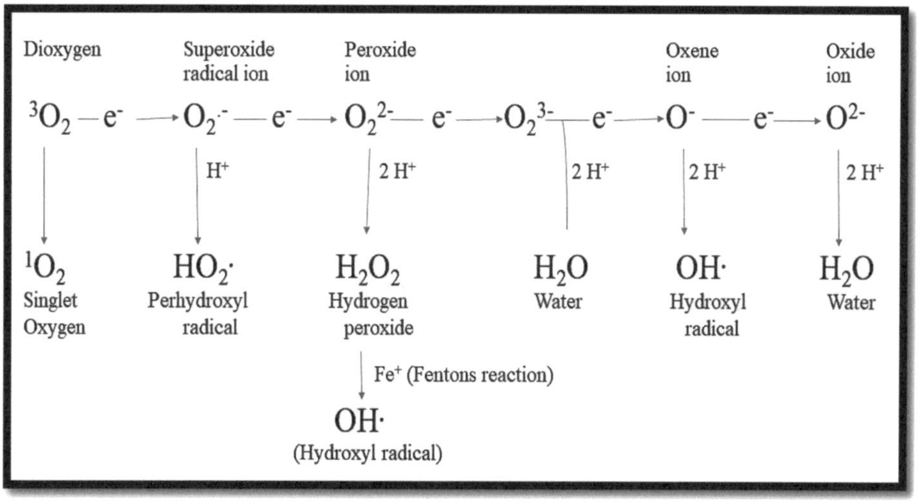

Figure 5.
Production of various ROS by energy transfer (or) sequential univalent reduction of ground state triplet oxygen.

oxygen ($^1O_2$) and hydroxyl radical ($OH^{\bullet}$) are mainly produced by enzymatic or non-enzymatic processes during photosynthesis (**Figure 5**). Their production occurs also in components of electron transport system in the mitochondria by partial reduction or oxidation of atmospheric oxygen [53]. In some current studies, it has been shown that ROS have dual role in plant biology; involvement in vital signaling processes and as toxic by-products of aerobic metabolism [53].

4.2 Enzymatic and non-enzymatic antioxidants

There are several componentsutilized by plantsto cope up with oxidative stress, which are involved in ROS homeostasis modulation [54]. Plants produces various reactive oxygen species (ROS) continuously as bi-products of various metabolic pathways in different cellular compartments like chloroplast, mitochondria, and peroxisome. ROS have partially reduced forms of atmospheric oxygen and under normal conditions, their production in plant cells is balanced by their effective scavenging through enzymatic and non-enzymatic cascade (**Figure 6**). ROS can cause damage to different biomolecules namely DNA, proteins and lipids, and therefore by creating oxidative injury; it leads to a reduction in plant growth and development [56]. The equilibrium between the production and the scavenging of ROS may be perturbed by various stress factors. Thus, the disturbances of cellular homeostasis resulted in a sudden rise in intracellular levels of ROS leading to oxidative stress which in turn can cause substantial damage to cell structure and membrane integrity. To mask themselves from these toxic oxygen intermediates, plant cells contain both enzymatic and non-enzymatic components. Among them enzymatic antioxidantsare superoxide dismutase (SOD), catalase (CAT), ascorbate peroxidase (APX), dehydroascorbate reductase (DHAR), monodehydroascorbate reductase (MDHAR) and ascorbate (AsA), glutathione (GSH), carotenoids, glycine betaine, proline, α-tocopherol and flavonoids are the non-enzymatic antioxidants [51, 57]. Hence, stress induced oxidative damage of ROSs can only be counteracted by increased level of enzymatic and nonenzymatic antioxidants [54].

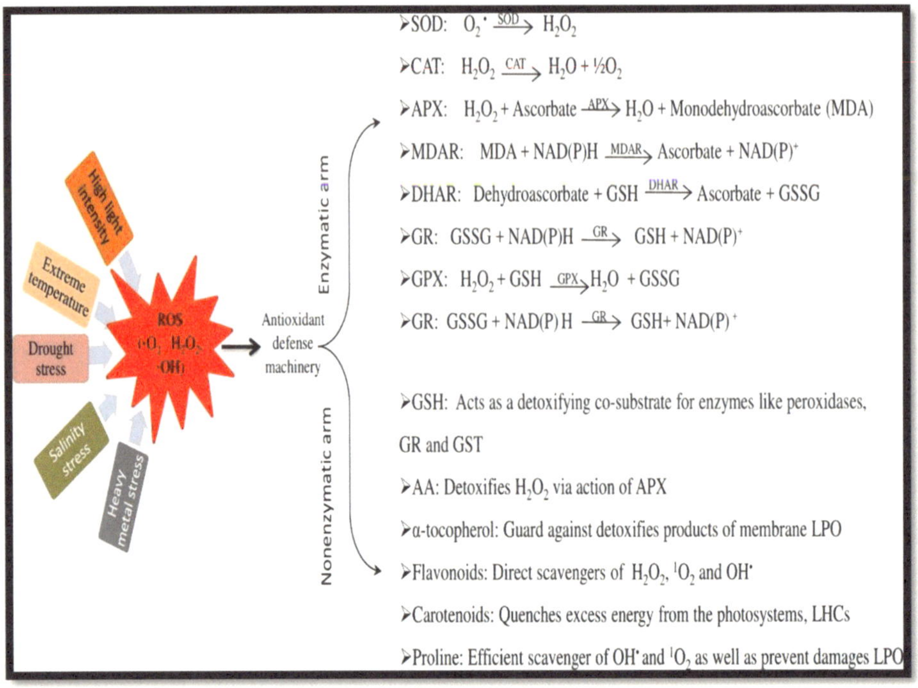

Figure 6.
ROS scavenging mechanism by antioxidant defense system in different stress conditions. Source: [55].

5. Molecular and genomic prospects for improvement of drought tolerance

Traditionally, there have been several efforts to develop drought-tolerant crop genotypes through usual breeding methods [58, 59]. In this method, two groups of plants with desirable traits are selected and crossed to obtain offsprings having new genetic arrangements [60]. Drought resistance is directly or indirectly incorporated in the crop species via genetic variability of traits and thus selection in breeding is ought to be useful. Important traits to target in plant breeding might include water-extraction efficiency, water-use efficiency, conductance of water, osmo-elastic adjustments and leaf area modulation [15]. Genetic data improves the efficiency of the breeding method. Polymorphisms based on molecular markers that occur naturally in the DNA like restriction fragment length polymorphisms (RFLPs), sequence characteristic amplified regions (SCARs), random amplified polymorphic DNA (RAPDs), simple sequence repeats (SSRs), amplified fragment length polymorphism (AFLPs), and others have been effectively utilized. The use of plant breeding methods has an enormous potential to accelerate drought-tolerant plant production and help drought management assist these plants [15].

Marker assisted selection (MAS) and genomic selection (GS) are the two well versed approaches of genomic assisted breeding. For the first approach, foremost step is to identify the molecular markers linked to the trait of interest so that selection can be performed in breeding programs. However, GS depends on progress of selection models based on genetic markers present on the whole genome and selection of genome estimated breeding values (GEBVs) in breeding populations through phenotyping of "training population".

MAS utilizes molecular markers in identification of quantitative trait loci (QTL) or specific genes that are linked with the target trait and are used to identify the individual with desirable alleles (**Figure 7**) [61]. Through these methods, QTLs for the traits linked with drought resistance are identified in various crops i.e., rice, wheat, maize, sorghum, pearl millet, soybean and many other crops [62–67].

Genomic selection utilizes all the markers available for a population of GEBVs and GS models are used for selection of elite lines without phenotyping [61]. Contrary to MAS, the information about QTLs is not the prerequisite for GS [68]. However, GS requires denser marker data than MAS. GS is being applied for breeding in maize tolerant to drought by the international maize and wheat improvement center (CIMMYT) [69]. Research efforts through this approach are progressing in other crops i.e., sugarcane, legumes and wheat [70–72].

Many studies have elucidated molecular responses in plants related to drought-induced transcription signaling pathways. In recent times, various stress-responsive genes and transcription factors having potential to mitigate drought-induced oxidative stress have been identified [73]. The TFs operate specific interaction with the cis-elements present in the genes' promoter region and, stimulate the expression of stress-inducible genes of various signaling pathways upon binding [74, 75]. These TFs are categorized into different families based on their conserved motifs that code their DNA binding domain (DBD), viz., APETALA 2 (AP2)/ethylene-responsive element binding factor (ERF); dehydration-responsive element binding protein (DREB); no apical meristem/*Arabidopsis* transcription activation factor, cup-shaped cotyledon

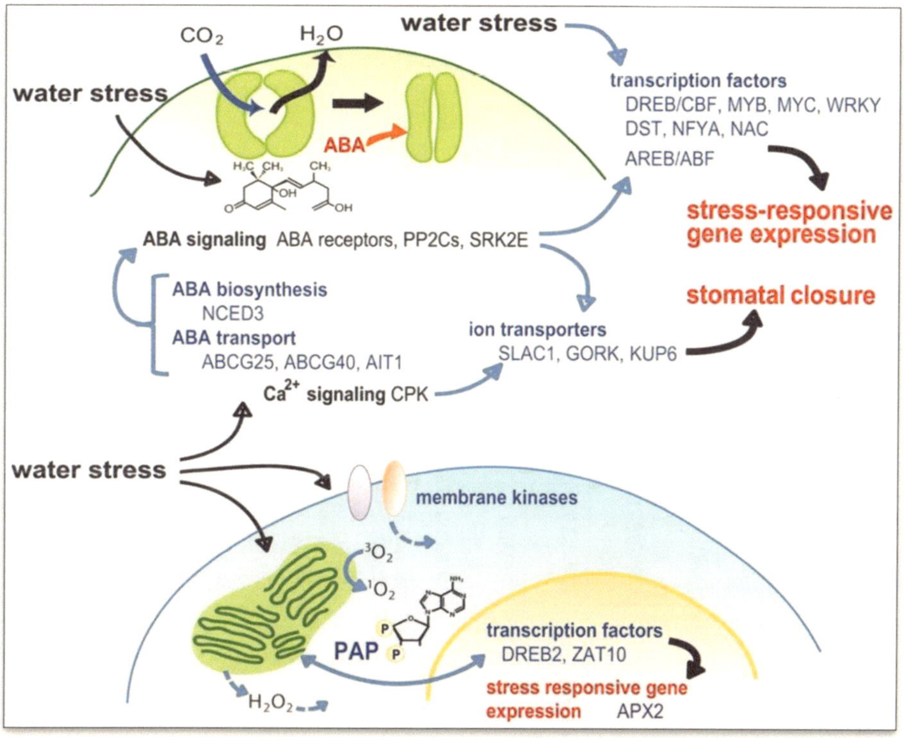

Figure 7.
Model for the role of signaling factors in stomatal closure and retrograde signaling during water stress. Source: [16].

Plant species	Genes	Pathway involved/activated	Function	References
Oryza sativa	*OsNAC5/6/9/10*	ABA responsive genes	drought avoidance	[80–83]
	NAC1/5/022		drought avoidance and activation of transcriptional regulation of various other genes	[84]
	bZIP23, ZFP252	Not identified	drought avoidance and activation of transcriptional regulation of various other genes	[85, 86]
	OsMYB2		drought avoidance and activation of transcriptional regulation of various other genes	[87]
	DREB1F/DREB2A/ EREBP1	mediates dehydration-inducible transcription	Enhanced ROS scavenging induced drought tolerance	[88]
Triticumaestivum	*TaNAC69, NAC2*	ABA responsive gene	drought avoidance and activation of transcriptional regulation of various other genes	[89, 90]
	TaPIMP1	Not identified	drought avoidance and activation of transcriptional regulation of various other genes	[91]
	DREB2/DREB3		Overexpression causes strengthening of the antioxidant defense system in response to drought stress	[92]
Arabidopsis thaliana	*NCED3*	ABA responsive gene	key enzyme of ABA biosynthesis	[93]
	AtABCG25		drought tolerance	[94]
	AHK1		positive regulator of osmosensing and drought tolerance	[95]
	OSCA1		membrane protein mediating osmotic stress responses	[96]
	Zat10	Not identified	drought avoidance and activation of transcriptional regulation of various other genes	[97]
	AREB1/ AREB2/ ABF3/ ABF4		Induce drought tolerance by trifurcating feed forward pathway	[98]
	bZIP28	stress sensor and transducer in ER stress signaling pathway	Activates brassinosteroid signaling and promotes acclimation to drought stress	[99]
Zea mays	*bZIP17*	noninducible expression of multiple genes involved in cell growth	Induced drought tolerance by promoting cell differentiation	[100, 101]

DOI: http://dx.doi.org/10.5772/intechopen.102780

Plant species	Genes	Pathway involved/activated	Function	References
Solanumtuberosum	*StMYB1R-1*	ABA responsive gene	drought avoidance and activation of transcriptional regulation of various other genes	[102]
Gossypiumhirsutum	*SnRK2*	ABA responsive gene	imparts cellular adaptation in response to dehydration stress.	[103]

Table 2.
Transcription factors involved in drought stress response in various crop species.

(NAC); related to abscisic acid insensitive (ABI3)/VIVIPAROUS 1 (VP1) (RAV); WRKY; auxin response factor (ARF); and SQUAMOSA-promoter binding protein (SBP). The DBDs of the AP2/ERF, DREB, NAC, SBP, and WRKY are named as per the names of their respective TFs family, whereas DBDs of ABI3/VP1 and ARF family of TFs are named as B3 family [76].

Biochemical and molecular factors involved in the induction of processes to alleviate the detrimental impacts of water stress include transcription, stress responsive genes like TaNAC69 (wheat), AP37 & OSNAC10 (rice), NF-YB2 (maize) and abscisic acid [16]. Transgenic expression of different stress responsive genes has been also utilized to confer increased tolerance to draught defecits. [77, 78]. The increased expression of these genes is frequently associated with a decreased plant growth rate and this could narrow down its practical use (**Table 2**) [79]. In this sense, genomic and related molecular tools could accentuate the genes that mitigate the stress effect so that efforts may help maintaining those genes in breeding programs [104]. Marker assisted breeding combined with traditional breeding as an integrated approach is the best approach for the improvement of the drought stress tolerance in plants. [105, 106].

6. Conclusion

Sustainable crop production to feed exponentially growing population is the major challenge to the scientific communities in the current global climate change scenario. Out of many productivity-limiting factors, drought stress is one of the most critical factor and of prime importance in the context of decreasing water availability for crop production. Water deficit leads to cellular damage and triggers an array of signaling pathways which in turn activate synthesis of gene transcripts associated with protective functions. In general, wilting occurs owing to physiological responses such as reduced turgor pressure, gaseous exchange, mineral assimilation and overall growth. The prominent result of these is reduced photosynthetic rate Many plant species are inherently equipped with drought tolerance mechanisms such as reduction in leaf area and canopy resistance. Both these mechanisms induce tolerance by cutting off excessive absorption of indecent light as a result of reduced surface area exposed to the incident radiations. In order to select for a tolerant genotype and/or traits conferring tolerance, robust phenotyping is a must. Marker assisted breeding to incorporate drought tolerance conferring quantitative trait loci (QTL) has proven to be effective and efficient. In addition, the knowledge generated by "OMICS" techniques (genomics, proteomics, transcriptomics, epigenomics and metabolomics) and transgenomics are potent and significant tools that would enable a researcher to develop an effective strategy for crop improvement programs in a less time-consuming cost-effective manner. So, an integrated approach will provide better understanding of mechanisms underlying drought stress and plants' response to it, and help in developing genotypes for dry environments in order to reduce the threat to global food security.

Acknowledgements

The author would like to thank to the co-authors for their valuable inputs.

Conflict of interest

The authors declare no conflict of interest.

Author details

Kousik Atta*, Aditya Pratap Singh, Saju Adhikary, Subhasis Mondal and Sujaya Dewanjee
Bidhan Chandra Krishi Viswavidyalaya, Mohanpur, Nadia, West Bengal, India

*Address all correspondence to: kousikatta1995@gmail.com

References

[1] Jaleel CA, Manivannan PA, Wahid A, Farooq M, Al-Juburi HJ, Somasundaram RA, et al. Drought stress in plants: A review on morphological characteristics and pigments composition. International Journal of Agriculture and Biology. 2009;**11**(1):100-105

[2] Farooq M, Wahid A, Kobayashi NS, Fujita DB, Basra SM. Plant drought stress: Effects, mechanisms and management. Sustainable Agriculture. 2009;**29**:153-188

[3] Kage H, Kochler M, Stützel H. Root growth and dry matter partitioning of cauliflower under drought stress conditions: Measurement and simulation. European Journal of Agronomy. 2004;**20**(4):379-394

[4] Lafitte HR, Yongsheng G, Yan S, Li ZK. Whole plant responses, key processes, and adaptation to drought stress: The case of rice. Journal of Experimental Botany. 2007;**58**(2):169-175

[5] Monneveux P, Sanchez C, Beck D, Edmeades GO. Drought tolerance improvement in tropical maize source populations: Evidence of progress. Crop Science. 2006;**46**(1):180-191

[6] Samarah NH. Effects of drought stress on growth and yield of barley. Agronomy for Sustainable Development. 2005;**25**(1):145-149

[7] Nayyar H, Kaur S, Singh S, Upadhyaya HD. Differential sensitivity of Desi (small-seeded) and Kabuli (large-seeded) chickpea genotypes to water stress during seed filling: Effects on accumulation of seed reserves and yield. Journal of the Science of Food and Agriculture. 2006;**86**(13):2076-2082

[8] Nam NH, Chauhan YS, Johansen C. Effect of timing of drought stress on growth and grain yield of extra-short-duration pigeonpea lines. The Journal of Agricultural Science. 2001;**136**(2): 179-189

[9] Samarah NH, Mullen RE, Cianzio SR, Scott P. Dehydrin-like proteins in soybean seeds in response to drought stress during seed filling. Crop Science. 2006;**46**(5):2141-2150

[10] Mazaheri LH, Nouri F, Zare AH. Effects of the reduction of drought stress using supplementary irrigation for sunflower (*Helianthus annuus*) in dry farming conditions., Pajouheshva-Sazandegi. Agronomy and Horticulture. 2003;**59**:81-86

[11] Kawakami J, Iwama K, Jitsuyama Y. Soil water stress and the growth and yield of potato plants grown from microtubers and conventional seed tubers. Field Crops Research. 2006;**95**(1):89-96

[12] Sinaki JM, Heravan EM, Rad AHS, Noormohammadi G, Zarei G. The effects of water deficit during growth stages of canola (*Brassica napus* L.). American-Eurasian Journal of Agricultural & Environmental Sciences. 2007;**2**:417-422

[13] Ogbonnaya CI, Sarr B, Brou C, Diouf O, Diop NN, Roy-Macauley H. Selection of cowpea genotypes in hydroponics, pots, and field for drought tolerance. Crop Science. 2003;**43**(3):1114-1120. DOI: 10.2135/cropsci2003.1114

[14] Rizza F, Badeck FW, Cattivelli L, Lidestri O, Di Fonzo N, Stanca AM. Use of a water stress index to identify barley genotypes adapted to rainfed and irrigated conditions. Crop Science.

2004;**44**(6):2127-2137. DOI: 10.2135/cropsci2004.2127

[15] Lisar SY, Motafakkerazad R, Hossain MM, Rahman IM. Causes, effects and responses. Water Stress. 2012;**25**(1):33

[16] Osakabe Y, Yamaguchi-Shinozaki K, Shinozaki K, Tran LS. ABA control of plant macro-element membrane transport systems in response to water deficit and high salinity. New Phytologist. 2014;**202**(1):35-49. DOI: 10.1111/nph.12613

[17] Abobatta WF. Drought adaptive mechanisms of plants—A review. Advances in Agriculture and Environmental Science. 2019;**2**(1):62-65. DOI: 10.30881/aaeoa.00021

[18] Lamaoui M, Jemo M, Datla R, Bekkaoui F. Heat and drought stresses in crops and approaches for their mitigation. Frontiers in Chemistry. 2018;**6**:26. DOI: 10.3389/fchem.2018.00026

[19] Earl HJ, Davis RF. Effect of drought stress on leaf and whole canopy radiation use efficiency and yield of maize. Agronomy Journal. 2003;**95**(3):688-696. DOI: 10.2134/agronj2003.6880

[20] Anjum SA, Xie XY, Wang LC, Saleem MF, Man C, Lei W. Morphological, physiological and biochemical responses of plants to drought stress. African Journal of Agricultural Research. 2011;**6**(9):2026-2032. DOI: 10.5897/AJAR10.027

[21] Gregorova Z, Kovacik J, Klejdus B, Maglovski M, Kuna R, Hauptvogel P, et al. Drought-induced responses of physiology, metabolites, and PR proteins in Triticumaestivum. Journal of Agricultural and Food Chemistry. 2015;**63**(37):8125-8133. DOI: 10.1021/acs.jafc.5b02951

[22] Tatar Ö, Gevrek MN. Lipid peroxidation and water content of wheat. Asian Journal of Plant Sciences. 2008;7:409-412

[23] Bennett D, Reynolds M, Mullan D, Izanloo A, Kuchel H, Langridge P, et al. Detection of two major grain yield QTL in bread wheat (Triticumaestivum L.) under heat, drought and high yield potential environments. Theoretical and Applied Genetics. 2012;**125**(7):1473-1485. DOI: 10.1007/s00122-012-1927-2

[24] Atta K, Chettri P, Pal AK. Physiological and biochemical changes under salinity and drought stress in ricebean [vignaumbellata (thunb.) ohwi and ohashi] seedlings. IJECC. 2020;**10**(8):58-64. DOI: 10.9734/ijecc/2020/v10i830218

[25] Alghabari F, Ihsan MZ, Hussain S, Aishia G, Daur I. Effect of Rht alleles on wheat grain yield and quality under high temperature and drought stress during booting and anthesis. Environmental Science and Pollution Research. 2015;**22**(20):15506-15515. DOI: 10.1007/s11356-015-4724-z

[26] Farooq M, Wahid A, Kobayashi NS, Fujita DB, Basra SM. Plant drought stress: Effects, mechanisms and management. Sustainable Agriculture. 2009;**29**:153-188. DOI: 10.1007/978-90-481-2666-8_12

[27] Abid M, Ali S, Qi LK, Zahoor R, Tian Z, Jiang D, et al. Physiological and biochemical changes during drought and recovery periods at tillering and jointing stages in wheat (Triticumaestivum L.). Scientific Reports. 2018;**8**(1):1-5. DOI: 10.1038/s41598-018-21441-7

[28] Singh J, Thakur JK. Photosynthesis and abiotic stress in plants. In: Biotic and Abiotic Stress Tolerance in Plants. Singapore: Springer; 2018. pp. 27-46. DOI: 10.1007/978-981-10-9029-5_2

[29] Thalmann M, Santelia D. Starch as a determinant of plant fitness under abiotic stress. New Phytologist. 2017;**214**(3):943-951. DOI: 10.1111/nph.14491

[30] Krasensky J, Jonak C. Drought, salt, and temperature stress-induced metabolic rearrangements and regulatory networks. Journal of Experimental Botany. 2012;**63**(4):1593-1608. DOI: 10.1093/jxb/err460

[31] Chen C, Xie Z, Liu X. Dynamic transformation of the substances of osmotic adjustment in winter wheat under iso-osmotic salt and drought stresses. Bulletin of Botanical Research. 2009;**29**(6):708-713

[32] Jyoti B, Yadav SK. Comparative study on biochemical parameters and antioxidant enzymes in a drought tolerant and a sensitive variety of horsegram (Macrotylomauniflorum) under drought stress. American Journal of Plant Physiology. 2012;**7**(1):17-29

[33] Petrović G, Jovičić D, Nikolić Z, Tamindžić G, Ignjatov M, Milošević D, et al. Comparative study of drought and salt stress effects on germination and seedling growth of pea. Genetika-Belgrade. 2016;**48**(1):373-381

[34] Babu K, Rosaiah G. A study on germination and seedling growth of Blcakgram (Vigna mungo L. Hepper) germplasm against Polyethylene glycol 6000 stress. IOSR Journal of Pharmacy and Biological Sciences (IOSR-JPBS). 2017;**12**:90-98

[35] Du W, Wang M, Fu S, Yu D. Mapping QTLs for seed yield and drought susceptibility index in soybean (Glycine max L.) across different environments. Journal of Genetics and Genomics. 2009;**36**(12):721-731. DOI: 10.1016/S1673-8527(08)60165-4

[36] Hall AE, Ismail AM, Ehlers JD, Marfo KO, Cisse N, Thiaw S, et al. Breeding cowpea for tolerance to temperature extremes and adaptation to drought. In: Challenges and Opportunities for Enhancing Sustainable Cowpea Production. Ibadan, Nigeria: International Institute of Tropical Agriculture; 2002. pp. 14-21

[37] Toker C, Canci H, Yildirim TO. Evaluation of perennial wild Cicer species for drought resistance. Genetic Resources and Crop Evolution. 2007;**54**(8):1781-1786. DOI: 10.1007/s10722-006-9197-y

[38] Batool T, Ali S, Seleiman MF, Naveed NH, Ali A, Ahmed K, et al. Plant growth promoting rhizobacteria alleviates drought stress in potato in response to suppressive oxidative stress and antioxidant enzymes activities. Scientific Reports. 2020;**10**(1):1-9. DOI: 10.1038/s41598-020-73489-z

[39] Galindo A, Collado-González J, Griñán I, Corell M, Centeno A, Martín-Palomo MJ, et al. Deficit irrigation and emerging fruit crops as a strategy to save water in Mediterranean semiarid agrosystems. Agricultural Water Management. 2018;**202**:311-324. DOI: 10.1016/j.agwat.2017.08.015

[40] Álvarez S, Rodríguez P, Broetto F, Sánchez-Blanco MJ. Long term responses and adaptive strategies of Pistacialentiscus under moderate and severe deficit irrigation and salinity: Osmotic and elastic adjustment, growth, ion uptake and photosynthetic activity. Agricultural Water Management. 2018;**202**:253-262. DOI: 10.1016/j.agwat.2018.01.006

[41] Tekle AT, Alemu MA. Drought tolerance mechanisms in field crops. World Journal of Biology and Medical Sciences. 2016;**3**(2):15-39

[42] Blum A. Plant water relations, plant stress and plant production. In: Plant Breeding for Water-Limited Environments. New York: Springer; 2011. pp. 11-52. DOI: 10.1007/978-1-4419-7491-4_2

[43] Dobra J, Motyka V, Dobrev P, Malbeck J, Prasil IT, Haisel D, et al. Comparison of hormonal responses to heat, drought and combined stress in tobacco plants with elevated proline content. Journal of Plant Physiology. 2010;**167**(16):1360-1370. DOI: 10.1016/j.jplph.2010.05.013

[44] Boulard T, Roy JC, Pouillard JB, Fatnassi H, Grisey A. Modelling of micrometeorology, canopy transpiration and photosynthesis in a closed greenhouse using computational fluid dynamics. Biosystems Engineering. 2017;**158**:110-133

[45] Wasaya A, Zhang X, Fang Q, Yan Z. Root phenotyping for drought tolerance: A review. Agronomy. 2018;**8**(11):241. DOI: 10.3390/agronomy8110241

[46] Zhang F, Wang P, Zou YN, Wu QS, Kuča K. Effects of mycorrhizal fungi on root-hair growth and hormone levels of taproot and lateral roots in trifoliate orange under drought stress. Archives of Agronomy and Soil Science. 2019;**65**(9):1316-1330. DOI: 10.1080/03650340.2018.1563780

[47] Tiwari P, Srivastava D, Chauhan AS, Indoliya Y, Singh PK, Tiwari S, et al. Root system architecture, physiological analysis and dynamic transcriptomics unravel the drought-responsive traits in rice genotypes. Ecotoxicology and Environmental Safety. 2021;**207**:111252. DOI: 10.1016/j.ecoenv.2020.111252

[48] Tzortzakis N, Chrysargyris A, Aziz A. Adaptive response of a native mediterranean grapevine cultivar upon short-term exposure to drought and heat stress in the context of climate change. Agronomy. 2020;**10**(2):249

[49] López-Galiano MJ, García-Robles I, González-Hernández AI, Camañes G, Vicedo B, Real MD, et al. Expression of miR159 is altered in tomato plants undergoing drought stress. Plants. 2019;**8**(7):201

[50] Qi J, Song CP, Wang B, Zhou J, Kangasjärvi J, Zhu JK, et al. Reactive oxygen species signaling and stomatal movement in plant responses to drought stress and pathogen attack. Journal of Integrative Plant Biology. 2018;**60**(9):805-826. DOI: 10.1111/jipb.12654

[51] Roychowdhury R, Khan MH, Choudhury S. Physiological and molecular responses for metalloid stress in rice—A Comprehensive Overview. Advances in Rice Research for Abiotic Stress Tolerance. 2019;**1**:341-369. DOI: 10.1016/B978-0-12-814332-2.00016-2

[52] Sapna H, Ashwini N, Ramesh S, Nataraja KN. Assessment of DNA methylation pattern under drought stress using methylation-sensitive randomly amplified polymorphism analysis in rice. Plant Genetic Resources. 2020;**18**(4):222-230. DOI: 10.1017/S1479262120000234

[53] Mittler R. ROS are good. Trends in Plant Science. 2017;**22**:11-19

[54] Rai KK, Rai N, Rai SP. Salicylic acid and nitric oxide alleviate high temperature induced oxidative damage in Lablab purpureus L plants by regulating bio-physical processes and DNA methylation. Plant Physiology and Biochemistry. 2018;**128**:72-88. DOI: 10.1016/j.plaphy.2018.04.023

[55] Ullah A, Sun H, Yang X, Zhang X. A novel cotton WRKY gene,

GhWRKY6-like, improves salt tolerance by activating the ABA signaling pathway and scavenging of reactive oxygen species. Physiologiaplantarum. 2018;**162**(4):439-454. DOI: 10.1111/ppl.12651

[56] Hernández-Jiménez MJ, Lucas MM, de Felipe MR. Antioxidant defence and damage in senescing lupin nodules. Plant Physiology and Biochemistry. 2002;**40**(6-8):645-657. DOI: 10.1016/S0981-9428(02)01422-5

[57] Hasanuzzaman MI, Roychowdhury RA, Karmakar JO, Dey NA, Nahar KA, Fujita MA. Recent advances in biotechnology and genomic approaches for abiotic stress tolerance in crop plants. In: Genomics and Proteomics: Concepts, Technologies and Applications. Burlington, Canada: Apple Academic Press; 2015. pp. 333-366

[58] Nezhadahmadi A, Prodhan ZH, Faruq G. Drought tolerance in wheat. The Scientific World Journal. 2013;**2013**:12. DOI: 10.1155/2013/610721

[59] Rana RM, Rehman SU, Ahmed J, Bilal M. A comprehensive overview of recent advances in drought stress tolerance research in wheat (Triticumaestivum L.). Asian Journal of Agriculture and Biology. 2013;**1**(1): 29-37

[60] Khan MA, Iqbal M, Jameel M, Nazeer W, Shakir S, Aslam MT, et al. Potentials of molecular based breeding to enhance drought tolerance in wheat (Triticumaestivum L.). African Journal of Biotechnology. 2011;**10**(55):11340-11344

[61] Varshney RK, Terauchi R, McCouch SR. Harvesting the promising fruits of genomics: Applying genome sequencing technologies to crop breeding. PLoS Biology. 2014;**12**(6):e1001883. DOI: 10.1371/journal.pbio.1001883. 100

[62] Morris GP, Ramu P, Deshpande SP, Hash CT, Shah T, Upadhyaya HD, et al. Population genomic and genome-wide association studies of agroclimatic traits in sorghum. Proceedings of the National Academy of Sciences. 2013;**110**(2):453-458. DOI: 10.1073/pnas.1215985110.104

[63] Kollers S, Rodemann B, Ling J, Korzun V, Ebmeyer E, Argillier O, et al. Whole genome association mapping of Fusarium head blight resistance in European winter wheat (Triticumaestivum L.). PLoS One. 2013;**8**(2):e57500. DOI: 10.1371/journal.pone.0057500

[64] Brown PJ, Upadyayula N, Mahone GS, Tian F, Bradbury PJ, Myles S, et al. Distinct genetic architectures for male and female inflorescence traits of maize. PLoS Genetics. 2011;**7**(11):e1002383. DOI: 10.1371/journal.pgen.1002383. 102

[65] Huang X, Wei X, Sang T, Zhao Q, Feng Q, Zhao Y, et al. Genome-wide association studies of 14 agronomic traits in rice landraces. Nature Genetics. 2010;**42**(11):961-967. DOI: 10.1038/ng.695. 103

[66] Bidinger FR, Nepolean T, Hash CT, Yadav RS, Howarth CJ. Quantitative trait loci for grain yield in pearl millet under variable postflowering moisture conditions. Crop Science. 2007;**47**(3):969-980. DOI: 10.2135/cropsci2006.07.0465. 106

[67] Wang B, Ekblom R, Bunikis I, Siitari H, Höglund J. Whole genome sequencing of the black grouse (Tetraotetrix): Reference guided assembly suggests faster-Z and MHC evolution. BMC Genomics.

2014;**15**(1):1-3. DOI: 10.1186/1471-2164-15-180. 105

[68] Nakaya A, Isobe SN. Will genomic selection be a practical method for plant breeding? Annals of Botany. 2012;**110**(6):1303-1316. DOI: 10.1093/aob/mcs109. 107

[69] Crossa J, Perez P, Hickey J, Burgueno J, Ornella L, Cerón-Rojas J, et al. Genomic prediction in CIMMYT maize and wheat breeding programs. Heredity. 2014;**112**(1):48-60. DOI: 10.1038/hdy.2013.16. 109

[70] Gouy M, Rousselle Y, Bastianelli D, Lecomte P, Bonnal L, Roques D, et al. Experimental assessment of the accuracy of genomic selection in sugarcane. Theoretical and Applied Genetics. 2013;**126**(10):2575-2586. DOI: 10.1007/s00122-013-2156-z. 110

[71] Varshney RK, Mohan SM, Gaur PM, Gangarao NV, Pandey MK, Bohra A, et al. Achievements and prospects of genomics-assisted breeding in three legume crops of the semi-arid tropics. Biotechnology Advances. 2013;**31**(8):1120-1134. DOI: 10.1016/j.biotechadv.2013.01.001. 111

[72] Rutkoski JE, Heffner EL, Sorrells ME. Genomic selection for durable stem rust resistance in wheat. Euphytica. 2011;**179**(1):161-173. DOI: 10.1007/s10681-010-0301-1. 112

[73] Kudo M, Kidokoro S, Yoshida T, Mizoi J, Todaka D, Fernie AR, et al. Double overexpression of DREB and PIF transcription factors improves drought stress tolerance and cell elongation in transgenic plants. Plant Biotechnology Journal. 2017;**15**(4):458-471. DOI: 10.1111/pbi.12644

[74] Anumalla M, Roychowdhury R, Geda CK, Bharathkumar S, Goutam KD, Mohandev TS. Mechanism of stress signal transduction and involvement of stress inducible transcription factors and genes in response to abiotic stresses in plant. International Journal of Recent Scientific Research. 2016;**7**(8):12754-12771

[75] Joshi R, Wani SH, Singh B, Bohra A, Dar ZA, Lone AA, et al. Transcription factors and plants response to drought stress: Current understanding and future directions. Frontiers in Plant Science. 2016;7:1029. DOI: 10.3389/fpls.2016.01029

[76] Kidokoro S, Watanabe K, Ohori T, Moriwaki T, Maruyama K, Mizoi J, et al. Soybean DREB 1/CBF-type transcription factors function in heat and drought as well as cold stress-responsive gene expression. The Plant Journal. 2015;**81**(3):505-518. DOI: 10.1111/tpj.12746

[77] Rai KK, Rai AC. Recent transgenic approaches for stress tolerance in crop plants. In: Sustainable Agriculture in the Era of Climate Change. Cham: Springer; 2020. pp. 533-556. DOI: 10.1007/978-3-030-45669-6_23

[78] Liu Y, Liu X, Wang X, Gao K, Qi W, Ren H, et al. Heterologous expression of heat stress-responsive AtPLC9 confers heat tolerance in transgenic rice. BMC Plant Biology. 2020;**20**(1):1-1. DOI: 10.1186/s12870-020-02709-5. 115

[79] Hussain HA, Hussain S, Khaliq A, Ashraf U, Anjum SA, Men S, et al. Chilling and drought stresses in crop plants: Implications, cross talk, and potential management opportunities. Frontiers in Plant Science. 2018;**9**:393. DOI: 10.3389/fpls.2018.00393. 116

[80] Nakashima K, Yamaguchi-Shinozaki K, Shinozaki K. The transcriptional regulatory network in

the drought response and its crosstalk in abiotic stress responses including drought, cold, and heat. Frontiers in Plant Science. 2014;**5**:170. DOI: 10.3389/fpls.2014.00170. 118

[81] Jeong JS, Kim YS, Baek KH, Jung H, Ha SH, Do Choi Y, et al. Root-specific expression of OsNAC10 improves drought tolerance and grain yield in rice under field drought conditions. Plant Physiology. 2010;**153**(1):185-197. DOI: 10.1104/pp.110.154773

[82] Redillas MC, Jeong JS, Kim YS, Jung H, Bang SW, Choi YD, et al. The overexpression of OsNAC9 alters the root architecture of rice plants enhancing drought resistance and grain yield under field conditions. Plant Biotechnology Journal. 2012;**10**(7):792-805. DOI: 10.1111/j.1467-7652.2012.00697.x

[83] Jeong JS, Kim YS, Redillas MC, Jang G, Jung H, Bang SW, et al. OsNAC5 overexpression enlarges root diameter in rice plants leading to enhanced drought tolerance and increased grain yield in the field. Plant Biotechnology Journal. 2013;**11**(1):101-114. DOI: 10.1111/pbi.12011

[84] Saad AS, Li X, Li HP, Huang T, Gao CS, Guo MW, et al. A rice stress-responsive NAC gene enhances tolerance of transgenic wheat to drought and salt stresses. Plant Science. 2013;**203**:33-40. DOI: 10.1016/j.plantsci.2012.12.016

[85] De Schutter K, Tsaneva M, Kulkarni SR, Rougé P, Vandepoele K, Van Damme EJ. Evolutionary relationships and expression analysis of EUL domain proteins in rice (Oryzasativa). Rice. 2017;**10**(1):1-9. DOI: 10.1186/s12284-017-0164-3

[86] Xu P, Chen F, Mannas JP, Feldman T, Sumner LW, Roossinck MJ. Virus infection improves drought tolerance. New Phytologist. 2008;**180**(4):911-921

[87] Yang A, Dai X, Zhang WH. A R2R3-type MYB gene, OsMYB2, is involved in salt, cold, and dehydration tolerance in rice. Journal of Experimental Botany. 2012;**63**(7):2541-2556. DOI: 10.1093/jxb/err431

[88] Xiong H, Yu J, Li J, Wang X, Liu P, Zhang H, et al. Natural variation of OsLG3 controls drought stress tolerance in rice by inducing ROS scavenging. Plant Physiology. 2017;**178**(1):451-467. DOI: 10.1104/pp.17.01492

[89] Xue GP, Way HM, Richardson T, Drenth J, Joyce PA, McIntyre CL. Overexpression of TaNAC69 leads to enhanced transcript levels of stress up-regulated genes and dehydration tolerance in bread wheat. Molecular Plant. 2011;**4**(4):697-712. DOI: 10.1093/mp/ssr013

[90] Gahlaut V, Jaiswal V, Kumar A, Gupta PK. Transcription factors involved in drought tolerance and their possible role in developing drought tolerant cultivars with emphasis on wheat (Triticumaestivum L.). Theoretical and Applied Genetics. 2016;**129**(11):2019-2042. DOI: 10.1007/s00122-016-2794-z

[91] Zhang Z, Liu X, Wang X, Zhou M, Zhou X, Ye X, et al. An R2R3 MYB transcription factor in wheat, Ta PIMP 1, mediates host resistance to Bipolarissorokiniana and drought stresses through regulation of defense-and stress-related genes. New Phytologist. 2012;**196**(4):1155-1170. DOI: 10.1111/j.1469-8137.2012.04353.x

[92] Wani SH, Tripathi P, Zaid A, Challa GS, Kumar A, Kumar V, et al. Transcriptional regulation

of osmotic stress tolerance in wheat (Triticumaestivum L.). Plant Molecular Biology. 2018;**97**(6):469-487. DOI: 10.1007/s11103-018-0761-6

[93] Takahashi T, Murano T, Ishikawa A. SOBIR1 and AGB1 independently contribute to nonhost resistance to Pyriculariaoryzae (syn. Magnaportheoryzae) in Arabidopsis thaliana. Bioscience, Biotechnology, and Biochemistry. 2018;**82**(11):1922-1930. DOI: 10.1080/09168451.2018.1498727

[94] Yoo CY, Pence HE, Jin JB, Miura K, Gosney MJ, Hasegawa PM, et al. The Arabidopsis GTL1 transcription factor regulates water use efficiency and drought tolerance by modulating stomatal density via transrepression of SDD1. The Plant Cell. 2010;**22**(12): 4128-4141

[95] Tran LS, Urao T, Qin F, Maruyama K, Kakimoto T, Shinozaki K, et al. Functional analysis of AHK1/ATHK1 and cytokinin receptor histidine kinases in response to abscisic acid, drought, and salt stress in Arabidopsis. Proceedings of the National Academy of Sciences. 2007;**104**(51):20623-20628. DOI: 10.1073/pnas.0706547105

[96] Yuan F, Yang H, Xue Y, Kong D, Ye R, Li C, et al. OSCA1 mediates osmotic-stress-evoked Ca 2+ increases vital for osmosensing in Arabidopsis. Nature. 2014;**514**(7522):367-371. DOI: 10.1038/nature13593

[97] Xiao BZ, Chen X, Xiang CB, Tang N, Zhang QF, Xiong LZ. Evaluation of seven function-known candidate genes for their effects on improving drought resistance of transgenic rice under field conditions. Molecular Plant. 2009;**2**(1):73-83. DOI: 10.1093/mp/ssn068

[98] Sakuraba Y, Kim YS, Han SH, Lee BD, Paek NC. The Arabidopsis transcription factor NAC016 promotes drought stress responses by repressing AREB1 transcription through a trifurcate feed-forward regulatory loop involving NAP. The Plant Cell. 2015;**27**(6):1771-1787. DOI: 10.1105/tpc.15.00222

[99] Kataoka R, Takahashi M, Suzuki N. Coordination between bZIP28 and HSFA2 in the regulation of heat response signals in Arabidopsis. Plant Signaling & Behavior. 2017;**12**(11):e1376159. DOI: 10.1080/15592324.2017.1376159

[100] Zhan J, Li G, Ryu CH, Ma C, Zhang S, Lloyd A, et al. Opaque-2 regulates a complex gene network associated with cell differentiation and storage functions of maize endosperm. The Plant Cell. 2018;**30**(10):2425-2446. DOI: 10.1105/tpc.18.00392

[101] Kim J-S, Yamaguchi-Shinozaki K, Shinozaki K. ER-anchored transcription factors bZIP17 and bZIP28 regulate root elongation. Plant Physiology. 2018;**176**(3):2221-2230. DOI: 10.1104/pp.17.01414

[102] Shin D, Moon SJ, Han S, Kim BG, Park SR, Lee SK, et al. Expression of StMYB1R-1, a novel potato single MYB-like domain transcription factor, increases drought tolerance. Plant Physiology. 2011;**155**(1):421-432. DOI: 10.1104/pp.110.163634

[103] Kulik A, Wawer I, Krzywińska E, Bucholc M, Dobrowolska G. SnRK2 protein kinases—Key regulators of plant response to abiotic stresses. Omics: A Journal of Integrative Biology. 2011;**15**(12):859-872. DOI: 10.1089/omi.2011.0091

[104] Medina S, Vicente R, Amador A, Araus JL. Interactive effects of elevated [CO_2] and water stress on physiological traits and gene expression during

vegetative growth in four durum wheat genotypes. Frontiers in Plant Science. 2016;**7**:1738. DOI: 10.3389/fpls.2016.01738. 117

[105] Bhatnagar-Mathur P, Vadez V, Sharma KK. Transgenic approaches for abiotic stress tolerance in plants: Retrospect and prospects. Plant Cell Reports. 2008;**27**(3):411-424. DOI: 10.1007/s00299-007-0474-9. 119

[106] Cho EK, Hong CB. Over-expression of tobacco NtHSP70-1 contributes to drought-stress tolerance in plants. Plant Cell Reports. 2006;**25**(4):349-358. DOI: 10.1007/s00299-005-0093-2. 120

Chapter 6

Physiological and Molecular Adaptation of Sugarcane under Drought vis-a-vis Root System Traits

Pooja Dhansu, Arun Kumar Raja, Krishnapriya Vengavasi, Ravinder Kumar, Adhini S. Pazhany, Ashwani Kumar, Naresh Kumar, Anita Mann and Shashi Kant Pandey

Abstract

Among various abiotic stresses, water is reported as a rare entity in many parts of the world. Decreased frequency of precipitation and global temperature rise will further aggravate the situation in future. Being C_4 plant, sugarcane requires generous water for the proper growth. Plant root system primarily supports above-ground growth by anchoring in the soil and facilitates water and nutrients uptake from the soil. The plasticity and dynamic nature of roots endow plants for the uptake of vital nutrients from the soil even under soil moisture conditions. In sugarcane, the major part of root system are generally observed in the upper soil layers, while limited water availability shifts the root growth towards the lower soil layer to sustained water uptake. In addition, root traits are directly related to physiological traits of the shoot to cope up with water limited situations via reduction in stomatal conductance and an upsurge in density and deep root traits, adaptations at biochemical and molecular level which includes osmotic adjustment and ROS detoxification. Under stressed conditions, these complex interactive systems adjust homeo-statically to minimize the adverse impacts of stress and sustain balanced metabolism. Therefore, the present chapter deals with physiological and biochemical traits along with root traits that helps for better productivity of sugarcane under water-limited conditions.

Keywords: sugarcane, root traits, drought, osmotic adjustment

1. Introduction

Global climate change along with abiotic stresses are one of the major constraints limiting factors to crop productivity that influences various agronomic characteristics, such as biomass and other growth traits, phenology, and yield-contributing traits, of various crops. Depletion of water resources, irregular rainfall pattern, impermanent periods of low water availability, moisture-holding capacity of the soil, water losses through evapotranspiration and poor groundwater quality pushing agriculture closer to the water scarcity situations. Generally, the damage to crop plants due drought is unpredictable, but plants experience drought when either the water supply to the roots is limited or the loss of water through transpiration is very high. Severe droughts cause a substantial decline in crop yields through negative impacts on plant growth, physiology, and reproduction. The plant response to drought varies from species to species and cultivars, phenological stages of the plant, and the duration of plant exposure to the stress. Under drought, along with nutrient and water relations, vital physiological traits viz. chlorophyll content, photosynthesis, stomatal conductance, chlorophyll fluorescence, assimilate partitioning, canopy temperature depression, membrane stability, impaired radiation use efficiency and reduced absorption of photosynthetically active radiations were seriously disrupted. At the same time, several biochemical and metabolic processes contributing to general growth and development were constrained along with the production of reactive oxygen species (ROS) that negatively affect cellular homeostasis, expression of genes, synthesis of hormones. Sugarcane, a C_4 plant, with a long life cycle is highly sensitive to water deficit and divided into four major phenophases, i.e., germination, formative/tillering, grand growth and maturity. Among these, the formative stage is considered as highly sensitive to drought stress [1, 2] as it required 550 mm of rainfall [3], causing a significant reduction in cane yield up to 50% [4]. Water is the major constituent of cane and approximately 2.97 lakh ha of cane area is prone to the drought that distributed cellular osmotic balance, decreased turgor, inhibited photosynthesis, inhibition of enzyme activities and cellular processes, root architecture and morphology and leading to a reduction in yield [5–7]. In point of these, this book chapter mainly focuses on the importance of physiological, biochemical, molecular traits along with root traits interventions related to drought for better management.

2. Effect of drought stress on sugarcane production

Sugarcane, the world's major industrial C_4 crop mainly grows in the tropic/subtropic regions and used for sugar and bioenergy. In India, the tropical regions (Maharashtra, Andhra Pradesh, Tamil Nadu, Karnataka, Gujarat, Madhya Pradesh, Goa, Pondicherry and Kerala) shared about 45% and 55% and the Sub-tropical region (Uttar Pradesh, Bihar, Haryana and Punjab) accounted for about 55% and 45% of the total sugarcane area and production in the country, respectively. Prevailing climatic factors particularly atmospheric CO_2, temperature, precipitation are the key factors for sugarcane production worldwide. Among these, the availability of water in addition to atmospheric CO_2 is considered to have been a major driving force for the evolution and the ecological success of C_4 plants [8, 9]. Drought or prolonged dry periods are limiting crop yields in tropical worldwide areas, including sugarcane production by 40–60% [10, 11]. Sugarcane is a globally important crop for food and industrial input production, accounts for 60% of world sugar output as well as bio-ethanol and energy generation [12]. Sugarcane is also known as water guzzler consumes about 22.5 million liters of water per hectare during its long growing cycle. In tropical India, the total water requirement of the crop for optimum growth varies from 2000 to 3000 mm inclusive of rainfall, while in Sub-tropical India, the water requirement is 1400–1800 mm. Drought severely depresses cane yield to the tune of 40–60%, whereas, the sucrose formation and sucrose recovery are reduced up to 5%.

The severe drought causes the complete failure of crop and sucrose recovery. India is the largest consumer and the second-largest producer of sugar in the world. The average annual production of sugarcane is around 35.5 crore tonnes which is used to produce 3 crore tonnes of sugar. In India, most of the country's irrigation facilities are utilized by paddy and sugarcane, depleting water availability for other crops. Pressure on water due to sugarcane cultivation in States like Maharashtra has become a serious concern, calling for more efficient and sustainable water use through alternative cropping pattern. This is especially important in regions where groundwater use has reached a critical and overexploited stage or where more than 50% surface water is used for irrigating sugarcane alone. Drought is a serious problem, but under production processing. There are various strategies to solve this obstacle, C_4 plants are often considered to have mastered the art of drought control particularly as they are able to maintain leaf photosynthesis with closed stomata. Generally, C_4 plants have high water use efficiencies (WUEs), and the presence of the CO_2-concentrating mechanisms makes C_4 photosynthesis more competitive in conditions that promote carbon loss through photorespiration, such as high temperatures, high light intensities, and decreased water availability [13]. C_4 photosynthesis is characterized by the presence of a metabolic CO_2 pump that concentrates CO_2 in the vicinity of the main enzyme of carbon dioxide fixation, ribulose-1,5-bisphosphate carboxylase/oxygenase [13, 14]. This confers several important advantages in terms of WUE because it allows high rates of photosynthesis to occur even when stomata are closed while limiting flux through the photo-respiratory pathway [14]. Therefore, an emergent choice for sustainable sugarcane production is the identification of water-efficient cultivars or providing water for irrigation.

3. Root Biology

Roots provide anchorage and facilitate the acquisition of water and nutrients from the soil, hence understanding the multi-faceted aspects of root biology are key to plant water management [15]. Plants exhibit a high degree of root plasticity and can transiently adapt to various environmental stresses. Sugarcane is a deep-rooted crop, owing to its long growth cycle. Sugarcane roots may be highly branched superficial roots, downward oriented buttress roots or deeply penetrating agglomerations of vertical roots known as rope roots. Sugarcane root systems may reach depths of about 1.5 to 6.0 m [16]. Nevertheless, drought stress usually leads to the formation of deeper root systems, which aid in extracting water from Sub-surface soil [17]. Sugarcane genotypes with drought-tolerant mechanisms cope better under water-limited environments by maintaining high water status and investing a higher proportion of assimilates for root growth under stress. Maintenance of high-water uptake under drought stress is also facilitated by improved root/shoot ratio in drought-tolerant cultivars [18]. Namwongsa et al. [19] observed that roots of sugarcane genotypes showed reduced growth in the upper soil layers (0 to 30 cm) in response to drought stress, whereas growth in the lower soil layers (below 30 cm) increased substantially. Under water-limited conditions, assimilated partitioning towards the roots area relatively higher than to shoots [20]. With the decrease in the soil moisture content, the roots alter their distribution patterns, proliferating more into deeper soil layers for extracting and engaging a larger soil volume for water uptake. As moisture at the soil surface and top soil profile is reduced under drought, deep roots take up water from the deeper profile [21]. Such deep root schemes are characteristic features under drought and is an important consequence to soil drying, allowing some roots to continue their lifecycle under stress. Hence, root architecture and distribution strongly depend upon the moisture content of various soil layers.

Jangpromma et al. [18] noticed that drought stress significantly reduced root length, root surface area, root volume and root dry weight, with negligible effect on the root/shoot ratio. Most of the root morphological traits could fully recover when plants were re-watered following drought stress during the formative phase. The unchanged root/shoot ratio might explain that sugarcane invested in roots under drought conditions to mine more water from deeper soil layers. Drought stress reduced root growth of sugarcane by 50 to 80% when the soil water status reduced to a water potential of −0.07 MPa [22]. Endres et al. [23] reported a drought-tolerant sugarcane genotype with higher root length density and better field performance under stress. Vantini et al. [24] observed differentially expressed genes between tolerant and sensitive sugarcane varieties across different time intervals (1, 3, 5, and 10 days after withholding water) in root tissues. At the beginning of the stress (1 and 3 days), genes encoding proteins with protection function (chaperones, heat shock proteins, antioxidant enzymes and protease inhibitor proteins) were induced in the tolerant variety. Gene encoding a protein involved in ABA-response, a trehalose-phosphatase synthase (enzyme involved in the synthesis of trehalose) and serine/threonine kinase receptors also showed higher expression in the tolerant variety, revealing differences between sugarcane genotypes for water stress protection and adaptive mechanisms. Hence, root systems are more important sink as compared to above ground organs under drought stress, especially during the active growth vegetative stages. Root growth might be indicative of a drought resistance mechanism under water-limited conditions. The positive relationship between root length and soil water content at the end of the drought period in 40 cm soil layers re-emphasizes the advantage of deeper roots for extracting water over extended drought periods. Nonetheless, the association between root length and physiological responses to plant water status are very complicated. Several root systems are considered to be essential in sustaining plant productivity under drought. Overall root growth, branching and distribution pattern is crucial to improve the acquisition of water and nutrients from the soil, and are positively associated with drought resistance and yield performance under stress.

4. Physiological responses of sugarcane under drought

4.1 Plant water relations

Plant–water relations under stress conditions explain how plants control or maintained the hydration of their cells up to an optimal level because it has important implications in the physiological and metabolic processes. It controls almost all metabolic activities within the cell which are dependent on the availability of sufficient amount of water present. Relative water content signifies as an indicator of plant water balance [25] and it indicates the level of cellular and tissues hydration which is imperative for the physio biochemical processes [26]. Plants under water deficit condition tends to have lesser RWC which triggers stomatal closure resulting in decreased CO_2 uptake [27]. Generally, the tolerant sugarcane genotype displays a higher RWC than the susceptible clones and Silva et al. [26] have described that the tolerant clones maintain better RWC (~87%) than the susceptible genotypes (~80%). Almeida et al.

[28] also reported a 50% decline in RWC particularly in RB 943365 (drought-sensitive sugarcane genotype) while the tolerant clone RB 72910 (drought-tolerant sugarcane genotype) remained constant at 86% RWC upon exposure to the water stress for 30 days. Medeiros et al. [29] witnessed significant decline in RWC with average values of 88.7 and 90.7% in the varieties RB 867515 and RB 962962 varieties owing to decrease of water in the soil, correspondingly.

Water potential (ψ_w) and Osmotic potential (ψ_s) are the physiological parameter used for recording the extent of stress level in plant. Water deficit induced decrease in leaf water potential led to interfere with plants' ability to extract water from the soil and maintain turgor [30]. Water potential controls stomatal conductance, which affects transpiration and photosynthesis, and affects root water uptake driven by the potential difference between leaf and soil water. Leaf water potential (ψ_w) was significantly reduced by 34.50% due to drought stress. However, the reduction in water potential was comparatively less in resistant genotypes *viz.*, Co 99,004 and Co 99,012 (26.96 and 30.15%) while the reduction was more than 50% in sensitive genotypes *viz.*, CoVc 93,136 and Co 99,014 [31]. The RB 72910 variety was able to maintain higher values Ψw when compared to RB 943365 variety under water deficit [28]. The water potential (ψ_w) was significantly reduced during stress treatment and values around 11-fold times smaller than those found in the control treatment, on average − 1.19 and − 0.78 MPa in the RB 867515 and the RB 962962 varieties, respectively. On rewatering treatment, a recovery of the plant water status was observed, as expected, with no difference among treatments, the average values of that treatment were − 0.16 MPa in the RB 867515 variety and -0.14 MPa in the RB 962962 variety [29]. Osmotic potential (ψ_s) is another physiological parameter used for recording the extent of stress level in plants. Reduction in osmotic potential of leaves has been observed under drought stress in sugarcane cultivar [32]. Basra et al. [33] studied the water relations in drought sensitive (BL-4) and drought tolerant (CP 43/33) sugarcane varieties by exposing to 200 mol m^{-3} mannitol solution and found that decline in turgor was faster in BL-4 than in CP 43/33 with time. Bulk leaf osmotic pressures and cell wall solutes were higher in BL-4 than in CP 43/33. Biancos et al. [34] observed the maximum difference in osmotic potential, between well-watered and stressed plants (i.e., the maximum osmotic adjustment) of about −0.5 and − 0.6 MPa for mature leaves and shoot tips in sugarcane, respectively. Pooja et al. [35] noted a significant decline in leaf RWC, leaf osmotic potential and leaf water potential in four sugarcane varieties exposed to drought stress. In another study, Pooja et al. [2] found that sugarcane clones Co 05011 (78.77%) and Co 0238 (76.88%) showed/maintained better RWC under stress conditions with a mean reduction of 16.70% over Co-canes RWC.

4.2 Growth analysis

The leaf area fairly gives a good idea of the photosynthetic capacity of the plant. The reduction in leaf area index (LAI) under water stress was due to reduced leaf area and a number of green leaves per stalk. Water stress had a greater effect on the expansive growth of lamina than its dry mass. Among different sugarcane varieties, water deficits reduced LAI and the biomass of sugarcane throughout the growing period. Inman-Bamber [36] observed that the specific leaf area (SLA) was lower in drought-treated plants as compared to control plants. da Silva & da Costa [37] has observed a significantly higher reduction in LAI under rainfed conditions than in irrigated conditions. Gomathi et al. [31] reported that the drought treatment caused an average reduction of 27.4 and 17.4% in leaf area and LAI, respectively in sugarcane. Varieties

Co 99,004 and Co 99,008 transpired less water and showed a relatively higher photosynthetic rate with significant improvement in growth attributes, viz., shoot leaf production and LAI. LAI showed significant variation among the genotypes and treatments indicating the sensitivity of these parameters to drought treatment. The genotypes Co 99,004, Co 99,012 and Co 99,006 recorded higher LAI of 3.53, 3.40 and 3.06, respectively even under drought conditions indicating their adaptability for leaf area production. Farooq et al. [38] reported a substantial reduction of LAI with increasing drought levels in a sugarcane cultivar.

It is well-established fact that the plant infrastructure is decided by the growth parameters such as, CGR, RGR and NAR. This concept not only involves the final crop yield and its components, but also probes into the physiological events that have occurred early in the growth stages causing variation in yield potential [39]. Singh & Singh [40] reported that 20 days and 40 days irrigation intervals in different sugarcane varieties caused significant loss in dry matter production *viz.,* RGR, NAR, and CGR during stress period when all other conditions were favorable for the remaining period of growth. Ramesh [41] evaluated four commercial sugarcane varieties (Co 8021, Co 419, Co 8208 and Co 6304) under three levels of drought stress (severe, moderate and control) during the formative phase (60–150 days after planting) and found a higher reduction in NAR, LAI, CGR especially under water-limited drought conditions. Farooq et al. [38] reported that NAR was lower at 80 and 60% irrigation coefficient as compared to 100% irrigation coefficient and variety NSG-59 showed higher NAR. Leaf area and cane length exhibited 37.3% and 26.53% reduction under drought conditions in comparison to the control [2].

4.3 Gas exchange traits

One of the more immediate responses of the drought stress is a reduction in the water potential of plant tissues leading to diminished stomatal aperture [42, 43] and consequent reduction in transpiration rate and photosynthesis, as well as longer-term responses such as growth inhibition, and accumulation of osmolytes [44]. In C_4 plants some evidences demonstrate that photosynthesis is highly sensitive to water deficit [45]. Moreover, these plants present low recovery capacity mainly when water deficit exceeds the plant recovery capacity limiting the photosynthesis metabolic pathways [46]. Sugarcane plants when subjected to decreased soil water content under moderate (42%) and severe stress (22%) caused changes in all photosynthetic apparatus, such as stomatal closure, reduction of transpiration and photosynthetic rate, as well as in RWC, photochemical efficiency of photosystem II (PS II), and increase in leaf temperature [47, 48]. The photosynthetic rate and stomatal conductance decreased significantly in drought-tolerant (SP 83–2847 and CTC 15) and sensitive sugarcane cultivars (SP 86–155), when submitted to water deficit however higher reduction percent, was recorded in sensitive cultivar [49]. Medeiros et al. [29] also reported that when young sugarcane plants of two varieties RB 867515 and RB 962962 were subjected to irrigation suspension until total stomata closure, and then rewatered, a significant reduction on stomatal conductance, transpiration rate, and net photosynthesis were observed. RB 867515 showed a faster stomatal closure while RB 962962 slowed the effects of drought on the gas exchanges parameters with a faster recovery after rewatering. Farooq et al. [38] that maximum water use efficiency was observed under 60% irrigation coefficient as compared to 80% and 100% irrigation coefficient and under 60% irrigation coefficient maximum water use efficiency was recorded in variety NSG followed by HSF-240. Water stress also caused a reduction in gas exchange traits and the associated pigments by 56.57% in stomatal conductance (gS), 56.55% in photosynthetic rate (pN), 38.21% in transpiration rate (E), 28.01% in internal CO2 (Ci) and 16.86% in the chlorophyll content [2]. Maximum water use efficiency (pN/E) under drought stress was recorded in Co 0238 (4.12) and Co 98,014 (3.93).

4.4 Osmotic adjustments

The cellular response to turgor reduction is an osmotic adjustment. The osmotic adjustment is achieved in these compartments by the accumulation of compatible osmolytes and osmoprotectants. Through the process of osmotic adjustment, higher plants can survive in dry and saline conditions. In this process, an accumulation of organic and inorganic solutes that reduce cellular osmotic potential and a reduction in the hydraulic conductivity of the membranes occurs, possibly by decreasing the number of water channels (aquaporins). Once the turgor is recovered, growth can be restored. The accumulation of compatible solutes is often regarded as a basic strategy for the protection and survival of plants under salt stress. Osmolytes are the organic compounds that play role in maintaining fluid balance as well as cell volume. In situations where increased external osmotic pressure tends to rupture the plant cells, certain osmotic channels are switched on to allow the efflux of certain osmolytes. As these osmolytes move outside, they carry water with themselves preventing the cell from bursting out. Sugars, alcohols, amino acids, polyols, tertiary and quaternary ammonium and sulphonium compounds are some examples of such osmolytes. A variety of compounds such as amino acids and amides (e.g., proline), ammonium compounds (e.g., betaine) and soluble carbohydrates act as compatible solutes. Proline, which is widely found in higher plants, accumulates in stressed plants in larger amounts than other amino acids [50]. Proline is a strong source to store carbon, nitrogen and a purifier of free radicals. Proline also maintains the structure of cell membrane and proteins [51, 52] and contributes to membrane stability [53–55]. It may also act as a signaling regulatory molecule able to activate multiple responses that are components of the adaptation process [56]. Boaretto et al. [57] reported that leaf proline content in IACSP 96–2042 sugarcane genotype was significantly increased about 2.3 to 2.7 times as compared to SP 87–365 under severe water-deficient conditions. Among the varieties, the differential accumulation of proline may be due to the response of a variety towards the environment [58]. The overproduction of proline may also mean a greater stress impact in Co 86,032 as compared to CoC 671, thus, rendering higher salt tolerance in CoC 671 than Co 86,032. Proline has also been reported to accumulate to maintain the osmotic potential of the plant cell under stress. Medeiros et al. [29] reported that free proline content was significantly increased in drought-affected sugarcane plants, i.e., 81.2% in RB 867515 variety and 72% in RB 962962 variety as compared to control plants. After rewatering, these values returned to normal levels.

Sugars were the main solutes that contributed to osmotic adjustment (OA) particularly in growing leaves. According to Zhou & Yu [59], these changes are related to the activation of responses to cope with this adverse environmental condition, to assist in the maintenance of cell water relations. The accumulation of soluble carbohydrates during water-deficient is considered a plant response to maintain hydration of the shoot and also protect enzyme and membrane system through the stabilization of proteins and lipids [60, 61]. Drought caused increases of soluble sugars content (SS) in sugarcane variety IACSP 96–2042 and IACSP 94–2094 sugarcane cultivar. On the other hand, water withholding increased non-structural carbohydrate content (NSCC) in IACSP 94–2094 and IACSP 96–2042. Under well-hydrated conditions, SP 87–365 had the highest NSCC when compared to the others, which did not vary due to water deficit [57]. Medeiros et al. [29] reported that soluble carbohydrate increased under water suppression, drought treatment increase was 51.2% in RB 867515 variety (sugarcane) and 28% in RB 962962. After rewatering, these values returned to normal levels, in the carbohydrates content of the RB 962962 variety, which did not differ from the water suppression and control treatment, such as in the RB 867515 variety that did not differ from the water suppression treatment. The alteration of protein synthesis or degradation is one of the fundamental metabolic processes that affect drought tolerance. Medeiros et al. [29] reported that amino acids and protein were significantly increased in drought-affected plants by 23.5 and 27% in sugarcane variety RB 867515 variety and 51.1 and 31.82% in variety RB 962962, respectively as compared to control plants. After rewatering, these values returned to normal levels. With other osmoregulatory, proteins were the major contributors to the osmoregulation of both varieties. Jangpromma et al. [62] reported accumulation of an 18 kDa protein was K86–161 sugarcane line which was subjected to progressive water stress for 20 days. Ngamhui et al. [63] reported a 16.9 kDa class 1 heat shock protein and two isoform elongation (EF-Tu) proteins, which are associated with heat tolerance under moisture stress in sugarcane variety Khon Kaen 3. Pooja et al. [35] observed approx. Two-fold increase in total soluble carbohydrates, four folds in proline and two-fold increase in lipid peroxidation under severe water stress conditions of 30% available soil moisture (ASM).

Potassium is a major ion which helps in the regulation of osmotic pressure, providing water maintenance at cells and plants textures, activation of various enzymes and coordination of opening or closing of stomata which may cause more air and plant evaporation. Ge et al. [64] observed that drought stress-induced sharp decreases in total K content and its uptake in maize organs at different developmental stages and, in particular, detrimentally affected the nutrient uptake capability of roots. Severe drought stress caused more deleterious effects than moderate drought stress on total K uptake by plant organs. Errabii et al. [65] reported that K content decreased in mannitol-induced stressed in sugarcane calli and it shows that inorganic solutes seemed to have no contribution in the osmotic adjustment in mannitol-induced stressed sugarcane calli. Such disruptions were due to the water outflow and the leakage of essential ions such as potassium and calcium content in sugarcane calli. Pooja et al. [7] observed significantly reduced K^+ content in leaves (2.93 to 1.83%) at ASM levels 30% and 40% as compared to 50% ASM level.

4.5 Antioxidant defense mechanism

Drought stress triggers cellular dehydration, accumulation of low molecular compounds (osmolytes) like glycine betaine, proline, sugar, alcohols, increased abscisic acid levels, increased expression of genes, excessive generation of reactive oxygen species (ROS) such as superoxide, hydrogen peroxide (H_2O_2) and hydroxyl radical affecting cellular structures and metabolism. Generation of ROS [superoxide radical ($O_2^{\bullet-}$), the hydroxyl radical (OH^-), hydrogen peroxide (H_2O_2) and singlet oxygen ($^1O_2$)] is observed as an outcome of the metabolic perturbations caused by osmotic effects of salt or dehydration stress as well as ionic toxicity of salt stress, particularly in chloroplast and mitochondria which ultimately leads to membrane leakage, lipid peroxidation, protein degradation and reduced enzyme activities.

Elevated production of ROS can seriously disrupt cellular homeostasis and normal metabolisms through oxidative damage to lipids, protein, and nucleic acid. Hydrogen peroxide is considered as one of the potential ROS which inhibits the functioning of the Calvin cycle. To mitigate the ROS-induced oxidative effects, plants have an antioxidant defense system that involves the generation of non-enzymatic and enzymatic antioxidants. Non-enzymatic antioxidants include phenolics, flavonoids, tocopherols, ASC, and GSH. Enzymatic antioxidants include superoxide dismutase (SOD), peroxidase (POX), catalase (CAT), as well as the enzymes of the ascorbate (ASC)–glutathione (GSH) cycle [GSH reductase (GR), ASC peroxidase (APX), monodehydroascorbate dehydrogenase (MDHAR), and dehydroascorbate reductase (DHAR)] that detoxify ROS [66–70].

Ascorbate peroxidase detoxifies hydrogen peroxide using ascorbate for reduction is present in chloroplasts, cytosol, mitochondria, apoplast and peroxisomes. By contrast, CAT is only present in peroxisomes, but it is indispensable for ROS detoxification during stress, when high levels of ROS are produced [71]. In addition, oxidative stress causes the proliferation of peroxisomes [72]. Catalase can be used to reduce hydrogen peroxide levels in the peroxisomes but it is absent in chloroplasts. The role of catalase is filled by specific ascorbate peroxidase. This peroxidase uses ascorbic acid as a hydrogen donor to break down hydrogen peroxide [73]. Water stress (PEG treatment) led to a significant increase in the activity of the antioxidant enzyme like CAT, POX, APX and SOD. Statistically, significant higher SOD activity was observed in salt (by 32%) or PEG (by 27%) stressed plants over the control in sugarcane callus of variety Co 86,032. CAT activity did not differ significantly in stressed and control plants [74]. Cia et al. [75] studied the antioxidant stress response of drought-tolerant (SP 832847 and SP 835073) and drought-sensitive (SP 903414 and SP 901638) sugarcane varieties to water deficit stress, which was imposed by withholding irrigation for 3, 10 and 20 days. SP 832847 exhibited higher CAT and APX activities than the other varieties in the early stage of drought, while the activities of GPOX and GR were the highest in the other varieties at the end of the drought stress period. Boaretto et al. [57] observed that the basal activity of CAT at 70% SAWC was greater in IACSP 95–5000 than in IACSP 94–2094. However, substantial increases in the total CAT activity were observed for both cultivars only at 30% SAWC. Ngamhui et al. [76] reported that under drought stress, tolerant sugarcane variety KK3 accounted 15% and 30% higher activity APX and POX, respectively as compared to variety SP72. Among three antioxidative enzymes, the highest activity of APX and POX was observed as compared to CAT in both the varieties. The activity of ROS content such as the superoxide radical, hydrogen peroxide and hydroxyl radical can cause oxidative stress and consequently membrane injury which leads to leakage of cellular content, peroxidation of membrane lipids, protein degrading, enzyme inactivation, pigment bleaching and disruption of DNA strands and thus cell death [76, 77]. Accumulation of hydrogen peroxide has not only negative consequences on living cells, but it is also involved in stress signaling and mediating the cellular redox status [78, 79]. Arora et al. [80] reported that as plants close the stomata under water deficit and reduce the internal CO_2 concentration, the generation of reactive oxygen species seems to stimulate mechanisms that reduce oxidative stress and so it may play an important role in drought tolerance.

Non-enzymatic antioxidant molecules can work synergistically with enzymatic ROS scavenging mechanisms to protect plant cells against oxidative damage. The non-enzymatic system is composed by ascorbic acid (AA), reduced glutathione (GSH), α-tocopherol, carotenoids, phenolics, flavonoids and proline [81, 82]. Proline (Pro) is an efficient scavenger of $OH^{\cdot}$ and $^1O_2$. Furthermore, Pro can function as a compatible osmolyte, molecular chaperone and carbon and nitrogen reserve and balances cytosolic pH [82, 83]. During water stress, Pro is accumulated in plants mainly due to increased synthesis and reduced degradation. Pro biosynthesis from glutamate is catalyzed by the enzymes Δ [1]-pyrroline-5-carboxylate (P5C) synthetase (P5CS) and P5C reductase (P5CR). Alternatively, Pro can be formed from ornithine that is converted into P5C/GSA *via* ornithine-δ-aminotransferase (OAT) [84, 85]. The observations recorded on antioxidative defense system have suggested possible key characteristics of drought tolerance and noted that low ASM levels induced the antioxidative defense system by increasing ROS and the specific activities of antioxidative enzymes, viz. peroxidase, catalase and ascorbate peroxidase [7]. The specific activity of these enzymes increased in varieties Co 0238 and CoS 767 at 60 and 90 DAP. Severe stress of 30% ASM levels also resulted in a sharp rise in total ascorbic acid content (9.36 to 13.14 mg/g), total soluble proteins (from 9.6 to 13.77 mg/g), and the increase was more in varieties Co 0238 and CoS 767.

5. Molecular responses of sugarcane under drought

During drought stress, the molecular responses include regulation of various signaling molecules, transcription factors (TFs) and drought induced genes (DIGs), which interact with each other and confer the drought tolerance potential to individual genotypes. Some of the major genes involved in drought tolerance are presented in **Table 1**. The genes are classified into two major categories: genes encoding functional proteins, and genes encoding regulatory proteins. Functional proteins play an important role in the protection of cells against dehydration, including late embryogenesis abundant (LEA) proteins, aquaporins (AQP), heat shock proteins (HSPs), ion transporters and metabolic enzymes. Regulatory proteins comprise calcium-binding proteins, protein kinases, transcription factors and signaling factors which can cause changes in plant physiology through signal transduction pathways, and regulate the expression of downstream genes [96]. LEA proteins exhibit important dehydrating protective functions during the late stage of embryo development in seeds. They play a key role in dehydration tolerance by capturing enough water into the cell. LEA proteins are composed of a high proportion of polar amino acids which makes them hydrophilic in nature, and can scavenge reactive oxygen species [96]. Transcription factors are one of the master regulators under stress, causing a significant change in gene expression. Manipulation of these regulatory elements may be beneficial for the enhancement of drought tolerance in sugarcane. It is now well established that both the transcription activators and repressors are involved in the drought stress tolerance [97, 98]. The major TF families involved in drought stress responses are WRKY, MYB, bZIP, NAC and DREB, which are perfect choices for genetic engineering to enhance stress tolerance. The most extensively studied TF family is WRKY which helps in regulating different physiological and metabolic processes [99]. The WRKY TF binds to the conserved DNA cis-element W-box to regulate further downstream processing of plant defense. Liu et al. [90] isolated and characterized the expression of WRKY TF of sugarcane using *E. coli* vector. It was observed that with the increase in the duration of stress, the relative expression level of WRKY increased, hence conferring drought tolerance in sugarcane.

To better understand the molecular basis of the physiological responses of sugarcane under stress conditions, high throughput gene expression studies have been conducted [48, 100–102]. Drought stress induces extensive signal transduction

S.No.	Genes	Predicted function	References
1.	LEA protein	Protection of macromolecules such as membranes	Liu et al. [86]
2.	Chaperones		Cagliari et al. [87]
3.	Dehydrin	Protection of cell against dehydration	Hayati et al. [88]
4.	Early responsive to dehydration protein (ERD)	Drought responsive genes	Devi et al. [89]
5.	WRKY	Drought induced Transcription factors	Liu et al. [90]
6.	ABRE-binding factor	Drought induced Transcription factors	Devi et al. [89]
7.	DRE-binding protein 2 (DREB2)	Drought induced Transcription factors	Reis et al. [91]
8.	NAC1 transcription factor	Drought induced Transcription factors	Devi et al. [89]
9.	Ethylene-responsive transcription factor (ERF3 gene)	Drought induced Transcription factors	Devi et al. [89]
10.	Trehalose 6-phosphate synthase	Signaling and trehalose metabolism	Hu et al. [92]
11.	Invertase	Plant development and gene expression regulation	Devi et al. [89]
12.	Sucrose phosphate synthase	ABA signaling and sucrose metabolism	Devi et al. [89]
13.	Calmodulin	Calcium binding protein and signaling molecule	Liu et al. [93]
14.	ABAR	ABA receptor	Devi et al. [89]
15.	P5CS	Proline metabolism and osmotic adjustment	Li et al. [94]
16.	BADH	Glycine betaine metabolism and osmotic adjustment	Chengmu et al. [95]

Table 1.
Important genes involved in the drought stress tolerance.

networks, comprising TFs, protein kinases and phosphatases [103–107]. As some of the plant responses under drought are ABA-dependent, both water stress and exogenous ABA treatment lead to several differentially expressed genes [108–110]. Free proline accumulation was correlated to drought stress tolerance in different sugarcane cultivars [111]. Transgenic sugarcane plants expressing a heterologous P5CS (encoding a proline biosynthetic enzyme), showed a positive correlation between enhanced proline content, increased biomass yield and photochemical efficiency of photosystem II under drought stress [112]. Iskandar et al. [113] indicated that a strong correlation between the expression of DIGs with increasing drought stress. Under severe drought stress, the expression of dehydrin proteins was induced [102], although there is no significant relation to sucrose accumulation [101, 113].

Author details

Pooja Dhansu[1*], Arun Kumar Raja[2], Krishnapriya Vengavasi[2], Ravinder Kumar[3], Adhini S. Pazhany[2], Ashwani Kumar[3], Naresh Kumar[3], Anita Mann[3] and Shashi Kant Pandey[1]

1 ICAR – Sugarcane Breeding Institute, Regional Center, Karnal, India

2 ICAR – Sugarcane Breeding Institute, Coimbatore, India

3 ICAR – Central Soil Salinity Research Institute, Karnal, India

*Address all correspondence to: poojadhansu@gmail.com

References

[1] Ramesh P, Mahadevaswamy M. Effect of formative phase drought on different classes of shoots, shoot mortality, cane attributes, yield and quality of four sugarcane cultivars. Journal of Agronomy and Crop Science. 2000;**185**:249-258

[2] Dhansu P, Kulshrestha N, Kumar R, Raja AK, Pandey SK, Goel V, et al. Identification of drought-tolerant co-canes based on physiological traits, yield attributes and drought tolerance indices. Sugar Tech. 2021;**23**:747-761. DOI: 10.1007/s 12355-021-00967-7

[3] Mall RK, Sonkar G, Bhatt D, Sharma NK, Baxla AK, Singh KK. Managing impact of extreme weather events in sugarcane in different agro-climatic zones of Uttar Pradesh. Mausam. 2016;**67**:233-250

[4] Mahadevaiah C, Hapase P, Sreenivasa V, et al. Delineation of genotype × environment interaction for identification of stable genotypes for tillering phase drought stress tolerance in sugarcane. Scientific Reports. 2021;**11**:18649. DOI: 10.1038/s41598-021-98002-y

[5] Pooja NAS, Chand M, Kumar A, Rani B, Kumari A, Kulshrestha N. Comparative evaluation of changes in protein profile of sugarcane varieties under different soil moisture regimes. International Journal of Current Microbiology and Applied Sciences. 2017;**6**:1203

[6] Pooja AS, Nandwal M, Chand A, Kumari B, Rani V, Goel S. Singh, Genotypic differences in growth behavior and quality parameters of sugarcane (Saccharum officinarum) varieties under moisture stress conditions. Indian Journal of Agricultural Sciences. 2019;**89**(1):65-72

[7] Pooja AS, Nandwal M, Chand A, Kumari B, Rani V, Goel N. Kulshreshtha, Soil moisture deficit induced changes in antioxidant defense mechanism of sugarcane varieties differing in maturity. Indian Journal of Agricultural Sciences. 2020;**90**(3):56-61

[8] Edwards EJ, Osborne CP, CAE S, Smith SA, C4 Grasses Consortium. The origins of C_4 grasslands: Integrating evolutionary and ecosystem science. Science. 2010;**328**:587-591

[9] Westhoff P, Gowik U. Evolution of C4photosynthesis–looking for the master switch. Plant Physiology. 2010;**154**:598-601

[10] Zhao D, Li Y-R. Climate change and sugarcane production: Potential impact and mitigation strategies. International Journal Agronomy. 2015;**547386**:1-10. DOI: 10.1155/2015/547386

[11] Joshi R, Wani SH, Singh B, Bohra A, Dar ZA, Lone AA, et al. Transcription factors and plants response to drought stress: current understanding and future directions Front. Plant Science. 2016;7:1029

[12] Waclawovsky AJ, Sato PM, Lembke CG, Moore PH, Souza GM. Sugarcane for bioenergy production: An assessment of yield andregulation of sucrose content. Plant Biotechnology Journal. 2010;**8**:263-276. DOI: 10.1111/j.1467-7652.2009.00491.x

[13] Edwards GE, Franceschi VR, Voznesenkaya EV. Single cell C4 photosynthesis versus the dual-cell (Kranz) paradigm. Annual Review of Plant Biology. 2004;**55**:173-196

[14] Edwards GE, Furbank RT, Hatch MD, Osmond CB. What does it

take to be C4? Lessons from the evolution of C4 photosynthesis. Plant Physiology. 2001;**125**:46-49

[15] Lynch JP. Root architecture and plant productivity. Plant Physiology. 1995;**109**:7-13

[16] Smith D, Inman-Bamber N, Thorburn P. Growth and function of the sugarcane root system. Field Crops Research. 2005;**92**:169-183. DOI: 10.1016/j.fcr.2005.01.017

[17] Jongrungklang N, Toomsan B, Vorasoot N, Jogloy S, Boote KJ, Hoogenboom GT, et al. Classification of root distribution patterns and their contributions to yield in peanut genotypes under mid-season drought stress. Field Crops Research. 2012;**127**:181-190

[18] Jangpromma N, Thammasirirak S, Jaisil P, Songsri P. Effects of drought and recovery from drought stress on above ground and root growth, and water use efficiency in sugarcane (*Saccharum officinarum L.*). Australian Journal of Crop Science. 2012;**6**(8):1298-1304

[19] Namwongsa J, Jongrungklang N, Songsri P. Genotypic variation in root distribution changes and physiological responses of sugarcane induced by drought stress. bioRxiv. 2018. DOI: 10.1101/503912

[20] Azhiri-Sigari T, Yamauchi A, Kamoshita A, Wade LJ. Genotypic variation in response of rain fed lowland rice to drought and rewatering II. Root growth. Plant Production Science. 2000;**3**:180-188

[21] Kavar T, Maras M, Kidric M, Sustar-Vozlic J, Meglic V. Identification of genes involved in the response of leaves of Phaseolus vulgaristo drought stress. Molecular Breeding. 2007;**21**:159-172

[22] Saliendra NZ, Meinzer FC. Symplast volume, turgor, stomatal conductance and growth in relation to osmotic and elastic adjustment in droughted sugarcane. Journal of Experimental Botany. 1991;**42**:1251-1259

[23] Endres L, Silva JV, Ferreira VM, Barbosa GVS. Photosynthesis and water relations in Brazilian sugarcane. The Open Agriculture Journal. 2010;**4**:31-37

[24] Vantini JS, Dedemo GC, Jovino Gimenez DF, Fonseca LF, Tezza RI, Mutton MA, et al. Differential gene expression in drought-tolerant sugarcane roots. Genetics and Molecular Research. 2015;**14**:7196-7207. DOI: 10.4238/2015. June.29.13

[25] Gonzalez L, Gonzalez-Vilar M. Determination of relative water content. In: Reigosa MJ, editor. Handbook of Plant Ecophysiology Techniques. Dordrecht: Kluwer Academic Publishers; 2001. pp. 207-212

[26] Silva MA, Jifon JL, Da Silva JAG, Sharma V. Use of physiological parameters as fast tools to screen for drought tolerance in sugarcane. Brazillian Journal of Plant Physiology. 2007;**19**:193-201

[27] Buckley TN. The control of stomata by water balance. New Phytologist. 2005;**168**(2):275-292

[28] Almeida CMA, Silva TD, Malafaia CB, Amaral DOJ, Arruda I, de Brito GG, et al. Proteomic and physiological analysis of response to water deficit in sugarcane. Wudpecker Journal of Agriculture Research. 2013;**2**(**1**):1-7

[29] Medeiros DB, da Silva EC, Nogueira RJMC, Teixeira MM, Buckeridge MS. Physiological limitations in two sugarcane varieties under water

suppression and after recovering. Theoretical and Experimental Plant Physiology. 2013;**25**(3):213-222

[30] Wahid A, Shabbir A. Induction of heat stress tolerance in barley seedlings by pre-sowing seed treatment with glycinebetaine. Plant Growth Regulation. 2005;**46**:133-141. DOI: 10.1007/s10725-005-8379-5

[31] Gomathi R, Vasantha S, Hemaprabha G, Alarmelu S, Shanthi RM. Evaluation of elite sugarcane clones for drought tolerance. Journal of Surgical Research. 2011;**1**:55-62

[32] Inman-Bamber NG, de Jager, 1.M. Effect of water stress on growth, leaf resistance and canopy temperature in field-grown sugar cane. Proceedings of South African Sugarcane Technology Association. 1986;**60**:156-161

[33] Basra SAM, Ahmed N, Khalid A. Water relations studies in water stresses sugarcane (*Saccaharum officinarum L.*). International Journal of Agriculture and Biology. 1999;**1**:1-4

[34] Biancos RL, Rieger M, Sung SS. Effect of drought on sorbitol and sucrose metabolism in sinks and sources of peach. Physiologia Plantarum. 2000;**108**:71-78

[35] Pooja AS, Nandwal M, Chand K, Singh AK, Mishra A, Kumar A, et al. Rani, Varietal variation in physiological and biochemical attributes of sugarcane varieties under different soil moisture regimes. Indian Journal of Experimental Biology. 2019b;**57**(10):721-732

[36] Inman-Bamber NG. Sugarcane water stress criteria for irrigation and drying off. Field Crops Research. 2004;**89**:107-122

[37] da Silva ALC, da Costa WAJM. Varietal variation in growth, physiology and yield of sugarcane under two contrasting water regimes. Tropical Agriculture Research. 2004;**16**:1-12

[38] Farooq U, Mehmood S, Afghan S, Shahzad A, Asad M. Comparative study on agro-physiology of sugarcane (*Saccharum officinarum L.*) genotypes at different irrigation co-efficient values. Pakistan Journal of Botany. 2015;**47**(2):527-532

[39] Patil RP, Chetti MB, Hiremath SM. Influence of agrochemicals on morpho-physiological characters, yield and yield components of sugarcane under moisture stress. Karnataka Journal of Agriculture Science. 2009;**22**(4):759-761

[40] Singh NB, Singh RG. Varietal differences in net assimilation rate and relative growth rate of sugarcane under moisture stress. Indian Journal of Plant Physiology. 1994;**37**:183-184

[41] Ramesh P. Effect of different levels of drought during the formative phase on growth parameters and its relationship with dry matter accumulation in sugarcane. Journal of Agronomy and Crop Science. 2000;**185**:83-89

[42] Meinzer FC, Grantz DA. Stomatal control of water use on sugarcane. Sugar Tech. 1989;**47**:72-75

[43] Yordanov I, Velikova V, Tsonev T. Plant responses to drought, acclimation, and stress tolerance. Photosynthetica. 2000;**38**:171-186

[44] Serraj R, Sinclair TR. Osmolyte accumulation: Can it really help increase crop yield under drought conditions? Plant, Cell & Environment. 2002;**25**:333-341

[45] Ghannoum O. C_4 photosynthesis and water stress. Annals of Botany. 2009;**103**(4):635-644

[46] Ripley B, Donald G, Osborne CP, Abraham T, Martin T. Experimental investigation of fire ecology in the C_3 and C_4 subspecies of Alloteropsis semialata. Journal of Ecology. 2010;**98**:1196-1203

[47] Rodrigues FA, Laia ML, Zingaretti SM. Analysis of gene expression profiles under water stress in tolerant and sensitive sugarcane plants. Plant Science. 2009;**176**:286-302

[48] Rodrigues FA, Graca JP, Laia ML, Nhani-Jr A, Galbiati JA, Ferro MIT, et al. Sugarcane genes differentially expressed during water deficit. Biologia Plantarum. 2011;**55**(1):43-53

[49] Graca JP, Rodrigues FA, Farias JRB, Oliveira MCN, Hoffmann-Campo CB, Zingaretti SM. Physiological parameters in sugarcane cultivars submitted to water deficit. Brazillian Journal of Plant Physiology. 2010;**22**:189-197

[50] Ghaderi N, Siosemardeh A. Response to drought stress of two strawberry cultivars (cv. Kurdistan and Selva). Horticulture and Environmental. Biotechnology. 2011;**52**(1):6-12

[51] Marandi RJ, Doostali PJ, Hassani R. Studying the tolerance of two apple roots to different concentrations of sodium chloride inside the glass. The Magazine of Horticulture Science in Iran. 2009;**40**(2):1-8

[52] Silva MA, Jifon JL, dos Santos CM, Jadoski CJ, da Silva JAG. Photosynthetic capacity and water use efficiency in sugarcane genotypes subject to water deficit during early growth phase. Brazillian Archives of Biology and Technology. 2013;**56**(5):735-748

[53] da Costa M, Huang BR. Osmotic adjustment associated with variation in bentgrass tolerance to drought stress. Journal of American Society of Horticulture Science. 2006;**131**(3):338-344

[54] Bandurska H, Gorny AG, Zielezinska M. Effects of water deficit on relative water content, proline accumulation and injury of cell membranes in leaves of old and modern cultivars of winter wheat. Zeszyty Problemowe Postepów Nauk Rolniczych. 2008;**524**:115-126

[55] Javadi T, Arzani K, Ebrahimzadeh H. Study of proline, soluble sugar, and chlorophyll a and b changes in nine Asian and one European pear cultivar under drought stress. Acta Horticulture. 2008;**769**:241-246

[56] Maggio A, Miyazaki S, Veronese P, Fujita T, Ibeas JI, Damsa B, et al. Does proline accumulation play an active role in stress-induced growth reduction? Plant Journal. 2002;**31**(6):699-712

[57] Boaretto LF, Carvalho G, Borgo L, Creste S, Landell MGA, Mazzafera P, et al. Water stress reveals differential antioxidante responses of tolerant and non-tolerant sugarcane genotypes. Plant Physiology and Biochemistry. 2014;**7**:165-175

[58] Hosseini T, Shekari F, Ghorbanli M. Effect of salt stress on ion content, proline and antioxidative enzymes of two safflower cultivars (*Carthamus tinctorius L.*). Journal of Food. Agriculture and Environment. 2010;**8**(2):1080-1086

[59] Zhou Q, Yu B. Plant physiology and biochemistry changes in content of free, conjugated and bound polyamines and osmotic adjustment in adaptation of vetiver grass to water deficit. Plant Physiology and Biochemistry. 2010;**48**:417-425

[60] Lawlor DW. Limitation to photosynthesis in water stressed leaves:

stomata versus metabolism and the role of ATP. Annals of Botany. 2002;**89**:1-15

[61] Yordanov I, Velikova V, Tsonev T. Plant responses to drought and stress tolerance. Bulgarian Journal of Plant Physiology. 2003;**38**(1):187-206

[62] Jangpromma N, Songsri P, Thammasirirak S, Jaisil P. Rapid assessment of chlorophyll content in sugarcane using a SPAD chlorophyll meter across different water stress conditions. Asian Journal of Plant Sciences. 2010;**96**:368-374

[63] Ngamhui NO, Akkasaeng C, Zhu YJ. Differentially expressed proteins in sugarcane leaves in response to water deficit stress. Plant Omics Journal. 2012;**5**(4):365-371

[64] Ge TD, Sun NB, Bai LD, Tong CL, Sui FG. Effects of drought stress on phosphorus and potassium uptake dynamics in summer maize (*Zea mays L.*) throughout the growth cycle. Acta Physiologiae Plantarum. 2012;**34**(6):2179-2186

[65] Errabii T, Gandonou CB, Essalmani H, Abrini J, Idaomar M, Senhaji NS. Growth, proline and ion accumulation in sugarcane callus cultures under drought-induced osmotic stress and its subsequent relief. African Journal of Biotechnology. 2006;**5**(16):1488-1493

[66] Hemaprabha G, Nagarajan R, Alarmelu S. Response of sugarcane genotypes to water deficit stress. Sugar Tech. 2004;**6**:165-168

[67] Jain R, Chandra A, Venugopalan VK, Solomon S. Physiological changes and expression of SOD and P5CS genes in response to water deficit in sugarcane. Sugar Tech. 2015;**17**:276-282. DOI: 10.1007/s12355-014-0317-2

[68] Sales C, Marchiori P, Machado R, Fontenele A, Machado E, Silveira J, et al. Photosynthetic and antioxidant responses to drought during sugarcane ripening. Photosynthetica. 2015;**53**:547-554. DOI: 10.1007/s11099-015-0146-x

[69] dos Santos CM, de Almeida Silva M, Lima GPP, d. Bortolheiro FPAP, Brunelli MC, Oliver R, et al. Physiological changes associated with antioxidant enzymes in response to sugarcane tolerance to water deficit and rehydration. Sugar Tech. 2015;**17**:291-304. DOI: 10.1007/s12355-014-0325-2

[70] dos Santos CM, de Almeida Silva M. Physiological and biochemical responses of sugarcane to oxidative stress induced by water deficit and paraquat. Acta Physiologiae Plantarum. 2015;**37**:1-14. DOI: 10.1007/s11738-015-1935-3

[71] Scandalios JG, Acevedo A, Ruzsa S. Catalase gene expression in response to chronic high temperature stress in maize. Plant Science. 2000;**156**:103-110

[72] Lee H, Xiong L, Ishitani M, Stevenson B, Zhu JK. Cold-regulated gene expression and freezing tolerance in an Arabidopsis thaliana mutant. Plant Journal. 1999;**17**:301-308

[73] Asada K. Production and action of active oxygen species in photosynthetic tissues. In: Foyer CH, Mullineaux PM, editors. Causes of photooxidative stress and amelioration of defense systems in plants. Boca Raton, FL: CRC Press; 1994. pp. 77-104

[74] Patade VY, Bhargava S, Suprasanna P. Salt and drought tolerance of sugarcane under iso-osmotic salt and water stress: growth, osmolytes accumulation, and antioxidant defense. Journal of Plant Interactions. 2011;**6**:275-282

[75] Cia MC, Guimaraes ACR, Medici LO, Chabregas SM, Azevedo RA. Antioxidant

response to water deficit by drought tolerant and sensitive sugarcane varieties. Annals of Applied Biology. 2012;**161**:313-324

[76] Ngamhui N, Tantisuwichwong RS, Zhu YJ, Li, XQ & Akkasaeng C. Relationship between drought tolerance with activities of antioxidant enzyme in sugarcane. Indian Journal of Plant Physiology. 2015;**20**:145-150

[77] Sairam RK, Tyagi A. Physiology and molecular biology of salinity stress tolerance in plants. Current Science. 2004;**86**:407-421

[78] Mittler R, Zilinskas B. Detection of ascorbate peroxidase activity in native gels by inhibition of the ascorbate-dependent reduction of nitroblue tetrazolium. Analytical Biochemistry. 1993;**212**:540-546

[79] Neill SJ, Desikan R, Clarke A, Hurst R, Hancock JT. Hydrogen peroxide and nitric oxide as signalling molecules in plants. Journal of Experimental Botany. 2001;**53**:1237-1247

[80] Arora A, Sairam RK, Srivastava GC. Oxidative stress and antioxidative systems in plants. Current Science. 2002;**82**:1227-1238

[81] Gill SS, Tuteja N. Reactive oxygen species and antioxidant machinery in abiotic stress tolerance in crop plants. Plant Physiology and Biochemistry. 2010;**48**:909-930

[82] Das K, Roychoudhury A. Reactive oxygen species (ROS) and response of antioxidants as ROS-scavengers during environmental stress in plants. Frontiers in Environmental Science. 2014;**2**:53. DOI: 10.3389/fenvs.2014.00053

[83] Verbruggen N, Hermans C. Proline accumulation in plants: a review. Amino Acids. 2008;**35**:753-759. DOI: 10.1007/s00726-008-0061-6

[84] Liang X, Zhang L, Natarajan SK, Becker DF. Proline mechanisms of stress survival. Antioxidants & Redox Signaling. 2013;**19**:998-1011. DOI: 10.1089/ars.2012.5074

[85] Bhaskara GB, Yang T-H, Verslues PE. Dynamic proline metabolism: Importance and regulation in water limited environments. Frontiers in Plant Science. 2015;**6**:484. DOI: 10.3389/fpls.2015.00484

[86] Liu J, Que Y, Guo J, Xu L, Wu J, Zheng Y, et al. Molecular cloning of sugarcane late embryogenesis abundant protein gene (LEA) and its expression character. Journal of Agricultural Biotechnology. 2009;**17**(5):836-842

[87] Cagliari TC, da Silva VC, Borges JC, Prando A, Tasic L, Ramos CH. Sugarcane Hsp101 is a hexameric chaperone that binds nucleotides. International Journal of Biological Macromolecules. 2011;**49**(5):1022-1030

[88] Hayati M, Sony S, Tati K, Amana DM. Isolation and characterization of dehydrin gene from sugarcane (*Saccharum officinarum L.*) involved in drought tolerance response. Menara Perkebunan. 2018;**86**(2):116-125

[89] Devi K, Prathima PT, Gomathi R, Manimekalai R, Lakshmi K, Selvi A. Gene expression profiling in sugarcane genotypes during drought stress and rehydration. Sugar Tech. 2019;**21**(5):717-733

[90] Liu J, Que Y, Guo J, Xu L, Wu J, Chen R. Molecular cloning and expression analysis of a WRKY transcription factor in sugarcane. African Journal of Biotechnology. 2012;**11**(24):6434-6444

[91] Reis RR, da Cunha BADB, Martins PK, Martins MTB, Alekcevetch JC, Chalfun-Júnior A, et al. Induced over-expression of AtDREB2A CA improves drought tolerance in sugarcane. Plant Science. 2014;**221**:59-68

[92] Hu X, Wu ZD, Luo ZY, Burner DM, Pan YB, Wu CW. Genome-wide analysis of the trehalose-6-phosphate synthase (TPS) gene family and expression profiling of ScTPS genes in sugarcane. Agronomy. 2020;**10**(7):969

[93] Liu J, Zhang C, Su W, Wu G, Fu X, Que Y, et al. Identification, Expression, and Functional Characterization of ScCaM in Response to Various Stresses in Sugarcane. Agronomy. 2021;**11**(11):2153

[94] Li J, Phan TT, Li YR, Xing YX, Yang LT. Isolation, transformation and overexpression of sugarcane SoP5CS gene for drought tolerance improvement. Sugar Tech. 2018;**20**(4):464-473

[95] Chengmu ZHZ. Physiological test of drought resistance and PCR amplification of BADH gene in sugarcane. Chinese Journal of Tropical Crops. 2003;**1**:59-62

[96] Yang X, Lu M, Wang Y, Wang Y, Liu Z, Chen S. Response Mechanism of Plants to Drought Stress. Horticulturae. 2021;7:50. DOI: 10.3390/horticulturae7030050

[97] Javed T, Shabbir R, Ali A, Afzal I, Zaheer U, Gao SJ. Transcription factors in plant stress responses: Challenges and potential for sugarcane improvement. Plants. 2020;**9**(4):491

[98] Ali A, Khan M, Sharif R, Mujtaba M, Gao SJ. Sugarcane Omics: An update on the current status of research and crop improvement. Plants. 2019;**8**(9):344

[99] Chen F, Hu Y, Vannozzi A, Wu K, Cai H, Qin Y, et al. The WRKY transcription factor family in model plants and crops. Critical Reviews in Plant Sciences. 2017;**36**:311-335

[100] Papini-Terzi FS, Rocha FR, Vencio RZN, Oliveira KC, Felix JD, Vicentini R, et al. Transcription profiling of signal transduction-related genes in sugarcane tissues. DNA Research. 2005;**12**:27-38. DOI: 10.1093/dnares/12.1.27

[101] Papini-Terzi FS, Rocha FR, Vêncio RZ, Felix JM, Branco DS, Waclawovsky AJ, et al. Sugarcane genes associated with sucrose content. BMC Genomics. 2009;**10**:120. DOI: 10.1186/1471-2164-10-120

[102] Rocha FR, Papini-Terzi FS, Nishiyama MY, Vêncio RZ, Vicentini R, Duarte RD, et al. Signal transduction-related responses to phytohormones and environmental challenges in sugarcane. BMC Genomics. 2007;**8**:71. DOI: 10.1186/1471-2164-8-71

[103] Zhu JK. Cell signaling under salt, water and cold stresses. Current Opinion in Plant Biology. 2001;**4**:401-406. DOI: 10.1016/S1369-5266(00)00192-8

[104] Zhu JK. Salt and drought stress signal transduction in plants. Annual Review of Plant Biology. 2002;**53**:247-273. DOI: 10.1146/annurev.arplant.53.091401.143329

[105] Singh K, Foley RC, O-ate-Sánchez L. Transcription factors in plant defense and stress responses. Current Opinion in Plant Biology. 2002;**5**:430-436. DOI: 10.1016/S1369-5266(02)00289-3

[106] Xiong L, Schumaker KS, Zhu JK. Cell signaling during cold, drought, and salt stress. Plant Cell. 2002;**14**(Suppl):S165-S183. DOI: 10.1105/tpc.000596

[107] Rabbani MA, Maruyama K, Abe H, Khan MA, Katsura K, Ito Y, et al. Monitoring expression profiles of rice genes under cold, drought, and high-salinity stresses and abscisic acid application using cDNA microarray and RNA gel-blot analyses. Plant Physiology. 2003;**133**:1755-1767. DOI: 10.1104/pp.103.025742

[108] Seki M, Ishida J, Narusaka M, Fujita M, Nanjo T, Umezawa T, et al. Monitoring the expression pattern of around 7,000 Arabidopsis genes under ABA treatments using a full-length cDNA microarray. Functional & Integrative Genomics. 2002;**2**:282-291. DOI: 10.1007/s10142-002-0070-6

[109] Himmelbach A, Yang Y, Grill E. Relay and control of abscisic acid signaling. Current Opinion in Plant Biology. 2003;**6**:470-479. DOI: 10.1016/S1369-5266(03)00090-6

[110] Shinozaki K, Yamaguchi-Shinozaki K. Gene networks involved in drought stress response and tolerance. Journal of Experimental Botany. 2007;**58**:221-227. DOI: 10.1093/jxb/erl164

[111] Guimarães ER, Mutton MA, Mutton MJR, Ferro MIT, Ravaneli GC, Silva JA, et al. Free proline accumulation in sugarcane under water restriction and spittlebug infestation. Science in Agriculture. 2008;**65**:628-633. DOI: 10.1590/S0103-90162008000600009

[112] Molinari HBC, Marur CJ, Daros E, De Campos MKF, De Carvalho JFRP, Pereira LFP, et al. Evaluation of the stress-inducible production of proline in transgenic sugarcane (*Saccharum* spp.): osmotic adjustment, chlorophyll fluorescence and oxidative stress. Physiologia Plantarum. 2007;**130**: 218-229. DOI: 10.1111/j.1399-3054.2007.00909.x

[113] Iskandar HM, Casu RE, Fletcher AT, Schmidt S, Xu J, Maclean DJ, et al. Identification of drought-response genes and a study of their expression during sucrose accumulation and water deficit in sugarcane culms. BMC Plant Biology. 2011;**11**:12. DOI: 10.1186/1471-2229-11-12

Chapter 7

Microgametophytic Selection as a Way to Improve Drought Tolerance in Cultivated Plants

Viktor Lyakh and Anatoliy Soroka

Abstract

In the current chapter, using different agricultural crops as an example, the effectiveness of pollen selection techniques for drought and heat resistance are demonstrated, as well as methods to evaluate plant drought tolerance by its male gametophyte. Germination of pollen from F_1 sunflower hybrid on the stigmas moistened with an osmotic resulted in drought resistance improvement of F_2 sporophytic generation, increasing the number of drought tolerant genotypes. Heating sunflower pollen increased both the plant adaptability to drought in dry field conditions and germination of seeds under osmotic stress. Pollination with heated pollen created opportunities to increase the share of drought resistant genotypes in the progeny of oil flax sporophytes. In interspecific tomato hybrids pollen treatment with high temperature was accompanied by a predominant elimination of unstable to various stresses pollen grains with cultivated species alleles in favor of wild ones. The high-temperature impact on the heterogeneous pollen population increased the proportion of drought-resistant genotypes in the sporophyte population and changed the Mendelian segregation ratios for a number of marker genes in maize. The genes were revealed, which influence drought resistance or are linked to the genes responsible for tolerance of pollen and plants to water stress in some crops.

Keywords: male gametophyte, osmotic, heating, pollen germination, selective elimination, sporophyte, drought tolerance, marker gene, segregation ratio, cultivated plant

1. Introduction

Gametophytic selection is the selection of genotypes in the sexual (haploid) generation of a plant's life cycle. Gametophytic selection is based on the different selective value of gametes, due to their genetic diversity. In this case, both micro and macrogametophytes can be subject to selection.

Selection in the haploid phase is widespread in higher plants. However, it was only recently shown that a significant part of the plant genome is expressed in the haploid phase and that a fairly large part of it is common for the sporophyte. The extensive overlap between the gametophytic and sporophytic phases constitutes the biological basis for the sporophyte response to gametophyte selection. Given that selection at the haploid level and, especially, in the male gametophytic generation, is capable to cause a significant shift in the genetic structure of populations in a short time, it can be actively involved in the breeding practice to search for valuable genotypes.

If each phase of a plant's life cycle had its own set of expressed genes, then selection at the gametophytic level would only lead to changes in the next gametophytic generation and would not affect the sporophyte. However, the results of numerous studies on the influence of gametophytic selection on sporophytic offspring [1–14] indicate that there is a significant degree of overlap between the two phases of the life cycle. It is believed that approximately 60% of the structural genes expressed in pollen are also expressed in the sporophyte [15–16].

Selection can operate in both the male and the female gametophytic generations. However, given the large size of the population of male gametes, the relative independence of pollen grains from the mother plant, the possibility of direct impact on them by environmental factors, the competition of many gametophytes within one style, male gametophytic selection, in comparison with female selection, is considered more effective. It is often compared to selection in microorganisms, given the large population size and the haploid state of the genome [3, 16, 17].

Considering the significant degree of genetic overlap of both phases of the plant life cycle, it is believed that the efficiency of gametophyte selection, especially pollen, may be higher than that of sporophyte [16]. This is explained by the fact that, firstly, the size of the population of male gametophytes is much larger than the size of the population of sporophytes. This makes it possible to apply high selection pressure. Secondly, it is much more difficult to obtain the desired gene combination at the sporophyte stage than at the gametophyte stage as the sporophyte of higher plants contains a double set of genes compared to the gametophyte. Because of this, the probability of selecting a complex combination of alleles is higher in the gametophytic generation. And thirdly, the haploid state of the gametophyte allows direct access to the recessive alleles.

2. Microgametophytic selection for drought tolerance in cultivated plants

2.1 Microgametophytic selection for drought tolerance in sunflower

Despite the fact that the sunflower is a mesophyte, it is very demanding to the level of available soil and air humidity. That is why the yield and efficiency of its growth are limited by moisture provided to the plants. In sunflowers the yield losses may reach 50% because of drought [18]. It should be noted that drought is usually accompanied by temperature increase which invokes damage to the plants and decreases the yield as well. Therefore, enhancing the drought and heat tolerance of the plants is an important direction of modern sunflower breeding.

2.1.1 Selection of drought tolerant genotypes during pollen germination on the stigmas in vivo

Undoubtedly, successful selection for any trait is possible only if there is a genetic variability for this trait. This also applies to the trait of drought tolerance in sunflowers. Therefore, in order to prove the effectiveness of microgametophytic selection for drought resistance, genetic variability for this trait at the level of the male gametophyte (pollen) was created artificially. For this, F_1 hybrid was obtained from the crossing of inbred sunflower lines with contrasting drought resistance which was determined by the relative length and dry weight of germinal roots after seed germination in sucrose solution. At the same time, it was assumed that those inbred lines were contrasting at the level of male gametophyte.

To carry out microgametophytic selection for drought resistance, a hybrid was taken obtained from crossing the morphological mutants of ZL 9 and ZL 102 sunflower lines developed at the Institute of Oilseed Crops of NAAS (Zaporozhye, Ukraine). In addition to different resistance to drought, one mutant line was characterized by chlorophyll deficiency of the "*virescent*" type at the top of the seedling, and the other one had leaves with fan venation. Both mutant traits are easily identified at the seedling stage. These inherited mutant traits were the result of our long-term research on chemical mutagenesis in sunflower [19, 20]. The presence of such "marker" traits in the parental components of the F_1 hybrid made it possible to unambiguously characterize the genetic structure of the F_2 segregating population.

The modified technique, proposed by Patil et al. for sorghum, was taken into account to select male gametophytes resistant to lack of moisture *in vivo* during pollen germination and pollen tube growth [21]. In the experiment, the stigmas of emasculated inflorescences were moistened with 10% of polyethylene glycol (PEG) 6000 solution. In this case, PEG 6000 solution was used as an osmotic selective barrier for pollen. After a little drying of the stigmas with PEG 6000 solution, pre-emasculated inflorescences of the F_1 hybrid were pollinated with fresh pollen. The inflorescences moistened with distilled water were used for pollination in the control. One cm^3 of pollen was taken for pollination of each head.

Table 1 shows the percentages of germination in 10% PEG 6000 solution of F_2 seeds [22], obtained during pollen germination of F_1 hybrid (ZL102 "*virescent*" x ZL9 "fan venation" crossing combination) in the presence and absence of osmotic on the stigmas of the hybrid plant.

As can be seen from the data presented in **Table 1**, the percentage of F_2 seed germination under conditions of osmotic stress was significantly higher in the experimental F_2 population, which was obtained after pollination with fresh pollen of the inflorescences moistened with 10% solution of PEG 6000. These results may indicate that such manipulation with a heterogeneous population of pollen from F_1 hybrid was effective and resulted in drought resistance improvement of F_2 sporophytic population, increasing there the number of drought-tolerant genotypes [23].

Table 2 shows the genetic structure of F_2 populations for "*virescent*" marker-trait. Change in the segregation for this marker-trait may indicate that the gene or genes, which determine the tested marker-trait, directly affect the drought tolerance of

Pollen treatment	F_2 seeds		Germination, %
	Total	Germinated	
Germination on the stigmas moistened with distilled water (control)	156	30	19.2
Germination on the stigmas moistened with 10% solution of PEG 6000	154	106	68.8***

*Note: ***differences from the control are significant at the 0.001 level of probability.*

Table 1.
Influence of pollen germination of F_1 hybrid in vivo in the presence of osmotic on germination of F_2 seeds in the medium with osmotic in sunflower.

Pollen treatment	F₂ phenotypes			Segregation Ratio
	Normal plants	Plants with marker-trait	Total	
Germination on the stigmas moistened with distilled water (control)	198	42	240	4.7:1
Germination on the stigmas moistened with PEG 6000	366	120	486	3.0:1**

Note: $\chi^2_{0.01}(df = 1) = 6.64$. ***differences from the control are significant at the 0.01 level of probability.*

Table 2.
Influence of pollen germination of F_1 sunflower hybrid in vivo in the presence of osmotic on segregation ratio in F_2 generation for "virescent" marker-trait.

plants or are linked to the genes, which determine the plant resistance to osmotic stress and dry environmental conditions [24].

The results of marker analysis for «*virescent*» trait showed the modification in segregation for the experimental F_2 population, which was obtained after germination of F_1 hybrid pollen on the stigmas moistened with osmotic, compared to the control. This modification resulted in a significant increase in the number of plants with the "*virescent*" trait. As for "fan venation", differences in segregation ratio for this trait were not revealed. Therefore, we can state the fact of the shift in the genetic structure of F_2 populations after carrying out gametophytic selection for drought tolerance which favored the survival of gametes with the "*virescent*" marker-trait and, at the same time, the elimination of gametes with an alternative trait.

In general, it must be admitted that pollination with a heterogeneous pollen population of stigmas moistened with the osmotically active substance can increase the drought resistance of F_2 sporophytic populations and modify their genetic structure for some marker traits. Also, it should be noted, that the simultaneous rise in the drought resistance and in the number of plants with the "*virescent*" marker-trait in F_2 population indicates that the gene or genes, which determine the "*virescent*" trait, directly influenced the drought resistance or were linked to the genes responsible for tolerance of pollen and plants to water stress.

2.1.2 Pollen heating as a way of selection drought tolerant genotypes

Treatment of pollen with high-temperature can also be an effective tool for breeders to overcome environmental stresses. Studies on the influence of F_1 pollen heating on the genetic structure of F_2 sporophytes showed that this technique can not only change the heat resistance of the segregating populations but its drought resistance as well. To estimate the effectiveness of pollen treatment for drought resistance improvement, comparisons were performed in the dry year 2013 on the number of plants that are able to bloom under such conditions. F_2 plants obtained from pollination with fresh or heated pollen from F_1 hybrids were analyzed.

Table 3 shows that in the year 2013 the amount of rainfall during the two months of vegetation before flowering was only about 60% of the long-term average. Before sowing in April and early May, there was almost no rainfall, while the average daily temperature was significantly higher than the average long-term data. This suggests that for cultivated sunflowers the environmental field conditions during the season of 2013 were arid, at the initial stages of growth and development especially.

Weather Parameter	April		May		June	
	Temperature, °C	Rainfall, mm	Temperature, °C	Rainfall, mm	Temperature, °C	Rainfall, mm
Average daily temperature	13.2	—	23.1	—	24.7	—
Total rainfall	—	8.0	—	29.0	—	31.5
Average long-term rates	8.5	35	16.0	40	19.4	62

Table 3.
Weather conditions in Zaporozhye region (Ukraine) before and after sowing of sunflower, 2013.

F_1 hybrids were grown in the field conditions during the year 2012 to obtain F_2 seeds by self-pollination. Before flowering, inflorescences of F_1 hybrids were isolated. One set of these plants was emasculated within 1–5 days for artificial pollination. The other set was grown without emasculation for collecting fresh pollen. Freshly collected pollen was immediately (in 5–10 min after collection) transferred to the laboratory, placed in parchment packets in a layer of 2–3 mm thick, and heated at the temperature of 60°C using an air bath oven for a period of 1 h and then used to self-pollinate the emasculated F_1 plants. At the same time, freshly collected pollen without temperature treatment was used to control pollinations. One cm^3 of pollen was taken for pollination of each head.

A viability test based on pollen germination on an artificial nutrient medium [25] demonstrates that the heating technique can significantly reduce the percentage of pollen germination on the artificial media. Thus, pollen treatment for 1 h at 60° C reduces the number of germinated pollen grains by almost 4 times, and treatment for 3 h—by more than 20 times, while the germination of pollen drops down to 1%.

Data on the influence of pollen heating in F_1 hybrids on the adaptability to drought in the F_2 resulting sporophytic offspring are presented in **Table 4**. As follows from the table, the number of flowering plants was significantly higher in F_2 populations obtained after 1-h pollen treatment than in the control. These facts indicate that gametophytic selection for heat tolerance increases the adaptability of the F_2 populations to drought stress [26].

In our earlier studies, we tested the selective effect of different high temperatures on the heterogeneous pollen populations of some interspecific sunflower hybrids [27]. The results showed that pollen heating at 40°C for 3 h did not cause the selective

Treatment	Number of F_2 seeds	Number of F_2 flowering plants	Frequency of F_2 plants, %
F_2 ZL9 mutant x ZL 95 mutant			
Fresh pollen	137	81	59.1 ± 4.20
Pollen heated at 60°C/1 h	540	395	73.1 ± 1.91**
F_2 ZL9 mutant x ZL 95 mutant			
Fresh pollen	250	167	66.8 ± 2.98
Pollen heated at 60°C/1 h	514	469	91.2 ± 1.25***

*Notes: **, *** the differences from the control are significant at the 0.01 and 0.001 levels of probability, respectively.*

Table 4.
Influence of pollen heating in F_1 sunflower hybrids on number of flowering plants in F_2 population.

Treatment	Number of F_2 seeds		Germination, %
	Total	Germinated	
F_2 ZL95 mutant × ZL9 mutant (15% sucrose as osmotic)			
Fresh pollen	480	43	9.0 ± 1.31
Pollen heated at 60°C/1 h	806	454	56.3 ± 1.75***
F_2 ZL102 mutant × ZL9 mutant (20% PEG 6000 as osmotic)			
Fresh pollen	156	30	19.2 ± 3.15
Pollen heated at 60°C/1 h	165	97	58.8 ± 3.83***

*Note: *** Differences from the control are significant at the 0.001 level of probability.*

Table 5.
Influence of pollen heating in F_1 hybrids on seed germination of the F_2 offspring in an osmotic solution in sunflower.

elimination of haploid genotypes. Only the use of a higher temperature for pollen treatment led to the shift in the genetic structure of the sporophytic population. As it turned out, pollen heating at the temperature of 60°C for 1 h was already effective.

Tolerance to drought of F_2 sporophytes after heating pollen of F_1 hybrids was also assessed by the seed germination in 15% sucrose or 20% PEG solutions. The F_2 seeds were germinated at the temperature of 25°C for a period of 4 days (sucrose) and 3 days (PEG 6000). After that, the percentage of seed germination was calculated **(Table 5)**.

As can be seen from **Table 5**, the heating of F_1 pollen increased the drought resistance of F_2 sporophytic population estimated by seed germination in a 15% sucrose solution or in 20% PEG 6000 solution. Germination in the osmotic of F_2 seeds that were developed after pollen heating was more than 3–6 times higher compared with the F_2 seeds resulting from the pollination of F_1 hybrids with fresh pollen.

These facts reveal that gametophytic selection for heat tolerance increases both the adaptability to the drought of the F_2 populations in dry field conditions and the germination of F_2 seeds in the conditions of osmotic stress. Thus, during the pollen heating, we observe the indirect selection that increases drought resistance of sporophytic offspring. It can be reasonably assumed that the genes which define sunflower drought resistance are linked to the genes determining heat resistance.

2.2 Microgametophytic selection for drought tolerance in oil flax

Under conditions of prolonged exposure to high temperatures and insufficient moisture supply, typical for the south of Ukraine, oil flax often reduces its potential productivity. The above makes the development of drought-resistant varieties of this crop urgent. The significant differences between flax genotypes insensitivity to elevated temperatures at the stage of the mature pollen grain and the discovered positive relationship between resistance to high temperatures of microgametophytes and drought tolerance of sporophytes, which we established, provide an opportunity for selection of drought-resistant genotypes at the microgametophytic level [10, 25].

Intervarietal F_1 hybrids K7487 × K7734 and K7734 × K7487 were used as the starting material, the parental lines of which significantly differed in their resistance to drought. Pollen was collected in the morning hours. Using a dissecting needle, the pollen from the anthers was shaken out onto a glass slide and after that, it was evenly distributed over the surface of the glass in a monolayer. Then glasses with pollen were

Pollen treatment	Germination, %	Root length, mm
BC_1 K7487 x (K7487 × K7734)		
Fresh pollen	5.0 ± 2.18	1.8 ± 0.40
Pollen heated at 35°C/1 h	19.7 ± 3.39***	1.7 ± 0.17
Pollen heated at 35°C/2 h	34.7 ± 2.57***	2.8 ± 0.28**
BC_1 K7487 × (K7734 × K7487)		
Fresh pollen	5.2 ± 2.11	1.2 ± 0.20
Pollen heated at 35°C/1 h	17.9 ± 3.18***	1.9 ± 0.25
Pollen heated at 35°C/2 h	24.2 ± 3.91***	5.7 ± 1.24**

*Note: **, *** Differences from the control are significant at the 0.01 and 0.001 levels of probability.*

Table 6.
Influence of pollen heating in F_1 hybrids on seed germination of BC_1 offspring on osmotic in oil flax.

placed in a thermostat and heated at 35°C for 1–2 hours. Pollen heating at a higher temperature is undesirable, since treatment at 45°C even for a short time results in a complete loss of pollen viability.

After heating, pollen was applied to the stigmas of one of the parental components of the hybrid. The fact of selection in the experiment was evaluated by the setting of seeds and capsules in comparison with pollination with freshly collected pollen. The selection efficiency was determined by the germinating BC_1 seeds in an osmotic medium (**Table 6**).

As a result of pollination with heated pollen, the setting of bolls and seeds significantly (almost by 2–3 times) decreased. Even after 1 h treatment of pollen, these parameters were significantly lower than in the control.

The heating of pollen from hybrid plants influenced both the percentage of seed germination and the length of roots developed in a medium with an osmotic. At the same time, the maximum differences from the control were observed when the pollen was heated for 120 min. Thus, the selection of pollen resistant to high temperatures in F_1 makes it possible to significantly increase the percentage of drought-resistant genotypes in the progeny of sporophytes obtained after pollination with the heated pollen. With the help of this methodological technique flax variety "Pivdenna nich" was developed which is a National Standard in Ukraine, and is perfectly adapted to the arid conditions of the south of the country.

2.3 Assessment of drought tolerance in castor bean by pollen

The climatic conditions of the south of Ukraine often have an adverse effect on the growth and development of castor bean plants, which are relatively unstable to drought and high temperatures. Therefore, the search for drought-resistant genotypes is the key aspect of plant breeding programs, which aim to obtain high-yielding varieties of this crop.

Express methods based on the analysis of pollen can be classified as promising in evaluating sporophytes for resistance to abiotic environmental factors, which is due not only to the experimentally revealed expression of a large number of genes at both stages of the plant life cycle but also to the ease of manipulation with such substance as pollen.

The correlation coefficient we established between the drought resistance of the sporophyte and the resistance of the male gametophyte to the action of high temperatures in the range of 0.72–0.89 opens the way to distinguish drought-resistant castor bean genotypes at the pollen level. An indicator of pollen resistance to elevated temperatures is the degree to which the percentage of pollen germination or the length of the pollen tube decreases after a stress factor is applied to mature pollen grains. When treating pollen of drought-resistant genotypes at the temperature of 40°C for 1 h, the degree of decrease in pollen germination percentage, as a rule, does not exceed 25%, while the length of a pollen tube may practically be of the same size.

As an example, data on the heat resistance of pollen can be provided for two genotypes of castor beans contrasting in resistance to drought (Khortytskaya 1 variety—tolerant to drought, K161 line—not tolerant to drought). Drought resistance of genotypes at the sporophyte stage was evaluated according to the method based on determining the osmotic pressure of cell sap by a refractometric method (−1.95 for Khortytskaya 1 variety, −1.0 for K161 line). In this case, the highest negative values of the water potential indicate the maximum ability of tissues to maintain it better, that is, to prevent the adverse effects of drought.

To estimate the tolerance of male gametophyte to high temperatures, pollen was subjected to temperature treatment at 40°C for 1 h, and then germinated *in vitro*. The tolerance of male gametophyte was assessed by the degree of decrease in pollen germination and pollen tube length in comparison with the control (pollen not subjected to heat processing).

As it turned out, in those genotypes whose sporophytes are characterized by high drought resistance, the degree of decrease in the germination of pollen after temperature treatment is minimal, while not drought-resistant lines show more significant inhibition of pollen germination when exposed to high temperatures (in Khortytskaya 1–18.8%, in K161 line—63.7%). The thermal treatment used also revealed significant differences between the accessions according to their ability to develop a long pollen tube. After heating pollen, in genotypes that were assessed as drought-resistant at the sporophyte stage, such stress caused a less significant decrease in the length of the pollen tubes than in not drought-resistant genotypes. In view of the above, a method of pollen evaluation can be proposed for screening drought-resistant castor bean samples.

2.4 Heat tolerance of tomato pollen and possibility of gametophytic selection for resistance to temperature stress

It is important in tomato breeding, especially for greenhouse cultivation, to keep in mind the heat resistance of pollen and to develop varieties and hybrids with this property. There are also known studies by G.I. Tarakanov et al. [28], who, taking into account the unequal value of different pollen grains in terms of heat resistance, used the most resistant gametes to obtain offspring. As a result of single and double selection, plants were raised that significantly exceeded the original specimen in terms of fruit set under conditions of elevated temperatures. Based on the available literature data, we can say that in tomatoes in the phase of mature pollen grain there is an opportunity for selection at the haploid level in order to change the tolerance to temperature stress of sporophytic offspring [28, 29].

Microgametophytic selection requires an answer to the question of the duration and magnitude of the effect on pollen in order to ensure both a sufficient strength of selection and the formation of an acceptable number of seeds in fruits. The material for the study was pollen of interspecific tomato hybrids obtained from the crossing of the Mo 500 mutant line with the wild species *Lycopersicum minutum* Rick and *Solanum pennellii* Corr. The pollen was heated in a thermostat at 58°C for 1, 2, 3, 6, and 12 h.

As a result of this experimental study, it was found that 6 and 12 h of heat treatment of pollen completely deprived it of the ability to germinate. Fruits were not set during pollination with such pollen, which indicated the loss of not only vitality but fertilizing ability as well. As optimal for gametophytic selection a 3 h treatment was accepted, which reduced the viability of pollen from 20–30% to 2–5%. The sum of temperatures under this effect amounted to 174°C and was close to the sum of lethal temperatures found for tomato varieties with high heat resistance of pollen.

Subsequently, the pollen of interspecific hybrids after 1 and 3 h of heating was used for pollination and raising F_2 and BC_1 offspring. It should be noted that the Mo 500 inbred line is a multimarket mutant, which is marked with 4 genes located in the 2nd and 6th chromosomes. These genes are easily identified in the early stages of plant development.

Marker analysis of F_2 and BC_1 segregating plant populations revealed differences after pollen treatment. Deviations from the control were insignificant for genes marking the second chromosome but significant for gene "*c*" (potato leaf), localized on the 6th chromosome (**Table 7**).

Only a 3 h treatment of F_1 hybrid pollen with high temperature led to a deviation of monogenic ratios in the direction of increasing the number of alleles of wild species. A decrease in the number of recessives at the *C* locus, emerged as a result of a 3 h temperature treatment of the hybrid pollen indicates that selection for heat resistance was accompanied by a predominant elimination of pollen grains of mutant type as compared to wild ones, such as *L. minutum* and *S. pennellii*.

In general, of the 4 studied loci, only the *C* locus has proved to be "temperature-sensitive" at the stage of mature pollen. Later, marker loci of the 4th and 11th chromosomes were involved in research. However, analysis of F_2 and BC_1 plants after heating of F_1 hybrid pollen did not reveal significant differences from the

Pollen treatment	F_2 phenotypes			Segregation Ratio
	Normal plants	Plants with marker-trait	Total	
F_2 Mo 500 × *L.minutum*				
Fresh pollen (control)	676	114	790	5.93:1
Pollen heated at 58°C/1 h	514	86	600	5.98:1
Pollen heated at 58°C/3 h	521	71	592	7.34:1*
F_2 Mo 500 × *S.pennellii*				
Fresh pollen (control)	342	84	426	4.07:1
Pollen heated at 58°C/1 h	348	84	432	4.14:1
Pollen heated at 58°C/3 h	370	70	440	5.29:1*
BC_1 Mo 500 (Mo 500 × *S.pennellii*)				
Fresh pollen (control)	140	161	301	0.87:1
Pollen heated at 58°C/3 h	196	154	350	1.27:1*

Note: $\chi^2_{0,05}$ *(df = 1) = 3.84; * differences from the control are significant at the 0.05 level of probability.*

Table 7.
Influence of pollen heating of F_1 tomato hybrids on the segregation ratio in F_2 or BC_1 generations for "potato leaf" marker-trait.

control in the allele frequencies of wild tomato species for the marker genes from those chromosomes.

It can be assumed that the 6th chromosome of a tomato or its section within the immediate vicinity of the *C* locus determines to some extent the sensitivity of mature pollen to high temperatures. It is important to note that the same locus was also "perceptive" to low temperatures as it fluctuated in segregations in accordance with pollen germination ability and pollen tube growth at cold stress.

To reveal the differences between the segregating populations of sporophytes obtained after pollination with fresh and heated pollen, those populations were tested against the high-temperature background at the stage of seedlings and 4–6 leaves. It was determined that in both cases the experimental populations exceeded the control ones in terms of heat resistance. That is, pollen selection for high-temperature resistance is effective and can be used as a helpful tool in breeding programs.

2.5 Heat tolerance of pollen and gametophytic selection for resistance to temperature stress in maize

Selection for sporophyte tolerance to high temperature and drought is a part of many maizes breeding programs. But there is little information on the heat sensitivity of reproductive processes and structures, specifically pollen. At the same time, it is known that maize pollen is characterized by a low ability to resist adverse environmental conditions. High temperature combined with low relative air humidity does have the most destructive effect.

2.5.1 Effects of high temperature on maize mature pollen grains

Pollen of maize inbred lines of various origins were collected on sunny days. Then it was placed on glass slides (without medium) in one pollen grain layer. Some slides were placed into a thermostat and heated in the dark at 35°C for 3, 5, 10, and 20 min and other slides were treated for 10, 20, and 30 min at 26°C. After the heat treatment, pollen was inoculated on the nutrient medium. Fresh pollen was used as the control. The percentage of pollen grain germination and pollen tube length were scored. Tolerance of pollen to high temperature was defined by the decrease in viability as compared to the control.

Effects of temperature on the percentage of pollen grain germination in two contrast maize inbreeds (A641 and MK386 lines) are presented in **Table 8**.

Heating mature pollen at 35°C decreased maize mature pollen grain viability even at very short exposures (3–5 min) in MK386 inbred but such treatments did not affect germination percentage in A641 inbred line. Longer treatments (10 min and more) considerably decreased pollen grain germination in A641 and completely inhibited the germination process in the MK386 line. In contrast to the MK386 line, the pollen of A641 inbred endured heating at 35°C for 20 min. Pollen of these accessions also lost its viability at 26°C. After 30 min treatment, MK386 pollen perished completely, while at the same exposure A641 pollen was characterized by the sufficiently large number of germinated pollen grains.

Differences in the tolerance of maize mature pollen to high temperatures were revealed among the inbreeds both at 26°C and 35°C. Even a 5-min treatment at 35°C resulted in decreases in pollen viability that varied in different inbreeds, as compared to the control, from 1.3% in A641 line to 99.1% in MK386 inbred. The effects of high temperatures on pollen tube length were similar. The data obtained allowed us to conclude that the genotypes with high-temperature resistant pollen were characterized

Heating temperature, °C	Heating exposition, min	A641	MK386
—	Control	45.9 ± 2.2	89.1 ± 1.3
26	10	45.6 ± 2.2	4.2 ± 0.8*
26	20	25.3 ± 1.8*	0*
26	30	18.2 ± 1.7*	0*
35	3	40.8 ± 2.1	9.3 ± 1.1*
35	5	45.3 ± 2.1	0.8 ± 0.3*
35	10	13.4 ± 1.5*	0*
35	20	1.0 ± 0.5*	0*

*Note: * differences from the control are significant at the 0.001 level of probability.*

Table 8.
Effects of mature pollen heating on pollen grain germination percentage in two maize accessions with contrasting heat resistance.

not only by higher germination percentage, but also a better ability to develop normal pollen tubes at high temperatures.

The mechanism which induces the rapid loss of viability of tricellulate cereal pollen is not yet clear. A reduction up to 80% of the original water content of maize pollen did not essentially affect its viability. With water loss greater than this value, pollen grains undergo irreversible changes.

A comparison of the heat resistance of pollen of inbred lines with their heat resistance at the stage of 20-day-old seedlings suggests a certain pattern. Lines with pollen more resistant to high temperature are also characterized by more resistant to this stress sporophyte. The correlation coefficient between the temperature resistance of the male gametophyte and the sporophyte turned out to be rather high and varied in different years from 0.6 to 0.78.

The positive relationship between the resistance to high temperatures of the gametophyte and sporophyte suggests that this trait is controlled by the same genetic system at both the haploid and diploid levels. In this case, the assessment of the sporophyte quality can be made based on the analysis of pollen traits.

Genotypic differences in the response of pollen to temperature stress give rise to the hope that the selection of mature pollen grains in a heterogeneous population will be successful in increasing the resistance of the sporophytic generation to high temperatures. In maize, mature pollen can be not only a tool for increasing sporophyte resistance. Mature pollen can be a breeding goal, given that the resistance of pollen grains to high temperatures after they are shed from the anthers is no less important for ensuring high crop yields than the resistance of a vegetative plant.

2.5.2 Effect of selection of pollen resistant to high temperature in F_1 on drought tolerance of F_2 sporophytes in maize

The effectiveness of gametophytic selection for drought resistance in maize can be demonstrated using the example of the VIR27 × MK01 hybrid. Its parental components were contrasting in resistance to high temperatures at the haploid and diploid stages of development.

To carry out the selection, the pollen of the hybrid was heated at the temperature of 35°C for 5–20 min. After heating, that pollen was used for self-pollination of the hybrid, having previously determined its viability on an artificial nutrient medium. For pollination, the pollen was used, which reduced its viability after heating by at least 80%. Pollination of each ear was carried out with a limited amount of pollen in order to exclude competition between haploid genotypes during pollen germination and pollen tube growth. Seeds (F_2 sporophytes) obtained after pollination of F_1 plant with fresh (control) and heated (treatment) pollen were sprouted in a solution with an osmotic (18 atm) and after a certain time, the percentage of germinated kernels was calculated.

As a result of heating the heterogeneous pollen population of this hybrid, the number of germinated seeds after 8 and 12 days under conditions of osmotic stress was 1.5 and 7.0 times higher than in the control. The data obtained allow to conclude about high efficiency of selection for drought resistance at the stage of mature pollen in maize.

High-temperature conditions when exposed to pollen are inevitably accompanied by low air humidity. Obviously, different reactions of haploid genotypes to high temperatures are due to different states of the membrane of pollen grains in genetically different haploid genotypes. Pollen grains resistant to these thermal treatments are characterized by such a membrane complex, which allows better water retention. As a result, such pollen retains its viability longer.

2.5.3 Changes in monogenic ratios for marker genes in F_2 as a result of high-temperature treatment of pollen from F_1 hybrids

A shift in Mendelian segregation ratios is one of the convincing evidence of the genetic activity of gametes. Segregation distortion for any marker gene as a result of treatment of a heterogeneous population of pollen is not yet evidence that this particular gene is expressed in the male gametophyte. A shift in the monogenic ratio for the marker locus can also be due to its linkage with the gene that determines the sensitivity of the gametophyte to the used selection agent. At the same time, conclusions about the localization of this gene (s) on a particular chromosome (chromosomes) of the genome can be drawn.

F_1 hybrids whose parent components were characterized by differences in heat tolerance at both gametophytic and sporophytic levels were used in our experiments. Mangelsdorf tester was one of the parents of those hybrids. This line is marked with 10 recessive genes, localized in different chromosomes. Seven markers were analyzed, which are typically expressed at the early stages of plant development: *bm2* (chromosome (Chr) 1), *lg1* (Chr 2), *a1* (Chr 3), *su1* (Chr 4), *gl1* (Chr 7), *j1* (Chr 8), *g1* (Chr 10).

Mature pollen grains of hybrids and pollen grains at the stage of their maturation were subjected to high-temperature treatment. In the latter case, maize tassels, which were transferred to a laboratory at the beginning of flowering, were exposed to temperature stress. After heat treatment, the pollen was used for self-pollination of the hybrids. In the control pollination was performed with fresh pollen.

With the help of marker analysis, the changes in segregation for seven recessive marker genes were evaluated in comparison with the control. For each F_2 population at least 400 plants were analyzed. In each case, the percentage of seed germination was quite high and amounted to more than 95%.

Pollen treatments of F_1 hybrids considerably influenced the monogenic ratio for some marker genes studied in the F_2 populations. The changes in dominant and recessive genotype ratios were found both after the treatment of mature pollen and pollen during its maturation. Following high-temperature treatment of pollen grains at the maturation stage, differences in the segregation for three markers loci—Gl_1, A_1, and

Male gametophyte stage	Temperature	
	Low	High
Pollen maturation	*a1* (3-th chromosome) *su1* (4-th chromosome) *g1* (10-th chromosome)	*a1* (3-th chromosome) *su1* (4-th chromosome) *gl1* (2-th chromosome)
Mature pollen grain	*bm2* (1-th chromosome) *a1* (3-th chromosome) *su1* (4-th chromosome) *g1* (10-th chromosome)	*bm2* (1-th chromosome) *a1* (3-th chromosome) *su1* (4-th chromosome) *g1* (10-th chromosome)
Pollen germination and pollen tube growth	*bm2* (1-th chromosome) *su1* (4-th chromosome) *lg1* (7-th chromosome) *j1* (8-th chromosome)	–

Table 9.
Marker genes for which a shift in monohybrid ratios in F_2 was found as a result of temperature treatment of male gametophyte of F_1 hybrids of maize.

Su_1 were observed. On the other hand, the treatment of mature pollen of F_1 hybrids led to a change in monogenic ratios for four marker loci on the first, third, fourth, and tenth chromosomes in comparison with the control (**Table 9**).

Comparing the data on changes in monohybrid ratios in F_2 as a result of exposure of the microgametophyte to high and low temperatures at the stages of pollen maturation, mature pollen grains and during germination and growth of the pollen tube, one can conclude that the direction of those changes largely depends on the stage of the male gametophyte than on the temperature. The male gametophyte of maize at each stage is characterized by its own set of expressed genes. Some of them, apparently, are expressed at the stages of development and functioning of the microgametophyte.

The presented data on the change in Mendelian segregation in F_2 under the influence of high temperature on the pollen of F_1 hybrids indicate the selectivity of this factor and the possibility of efficient selection of genotypes resistant to temperature stress at the microgametophytic level in maize.

3. Conclusions

Losses in crop yield caused by unfavorable abiotic factors, in particular drought, may reach large dimensions. Drought is usually accompanied by temperature spikes which invoke damage of plants and decrease the yield. Therefore, increasing the drought and heat tolerance are important directions in the modern breeding of cultivated plants.

Along with the traditional methods of selection for tolerance to abiotic stresses at the level of seeds, seedlings, or plants the methods of gametophytic selection, including pollen selection, can be successfully used. It was established that a large part of the genes expressed in the pollen are also exhibits in the plant. This fact indicates that gene transcription occurs in the haploid genome and therefore selection of the traits, which are under the control of genes expressed at both plant and microgametophyte (pollen) levels can be performed effectively.

In our research with sunflowers, when freshly collected pollen of F_1 plant was heated and used for self-pollination, the drought resistance of the next F_2 generation was significantly enhanced. This was confirmed by the large number of plants that can survive to bloom in dry field conditions as well as better F_2 seed germination in osmotic solutions. In another experiment, germination of sunflower pollen in vivo under osmotic stress with polyethylene glycol solution as the selective barrier increased seed germination on the osmotic medium in vitro.

High efficiency of the pollen selection technique for drought tolerance was revealed by us earlier in other crops, such as tomato, maize, castor bean, and linseed. It was in this way that the flax variety was obtained which is a National Standard in Ukraine, and is perfectly adapted to the dry conditions of the south of the country.

Evaluation at the level of pollen may also be of independent interest since the sensitivity to a drought of the plant reproductive system itself is a very important characteristic for many agricultural crops. A possibility to use this approach at the earliest stages of the breeding process, especially when the breeding material is quantitatively limited, is an undeniable advantage of the proposed assessment method.

In general, in the suggested chapter, using as an example different agricultural crop, the effectiveness of pollen selection techniques for drought resistance were demonstrated, as well as methods for assessing plant drought tolerance by its male gametophyte.

Author details

Viktor Lyakh[1]* and Anatoliy Soroka[2]

1 Zaporozhye National University, Ukraine

2 Institute of Oilseed Crops of the National Academy of Agricultural Sciences of Ukraine, Ukraine

*Address all correspondence to: lyakh@iname.com

References

[1] Chikkodi SB, Ravikumar RL. Influence of pollen selection for Alternaria helianthi resistance on the progeny performance against leaf blight in sunflower (Helianthus annuus L.). sex. Plant Reproduction. 2000;**12**(4):222-226. DOI: 10.1007/s004970050004

[2] Clarke HJ, Khan TN, Siddique KHM. Pollen selection for chilling tolerance at hybridisation leads to improved chickpea cultivars. Euphytica. 2004;**139**:65-74. DOI: 10.1007/s10681-004-2466-y

[3] Hormaza JI, Herrero M. Pollen selection. Theoretical and Applied Genetics. 1992;**83**:663-672. DOI: 10.1007/BF00226682

[4] Kravchenko AN, Lyakh VA, Toderash LG, Saltanovich TI, Paskal MK. Methods of Gamete and Zygote Selection in Tomatoes. Shtiintsa: Chisinau; 1988. p. 152 (In Russian)

[5] Lyakh VA, Soroka AI. Efficiency of microgametophytic selection for maize tolerance to temperature factor. Sel'skokhozyajstvennaya Biologiya. 1993;**3**:38-44

[6] Lyakh VA, Soroka AI. Pollen selection techniques for oilseed crop breeding. Scientific and Technical Bulletin of the Institute of Oilseed Crops NAAS. 2017;**24**:7-15. Available from: http://bulletin.imk.zp.ua/pdf/2017/24/Lyakh_24.pdf

[7] Lyakh VA, Soroka AI, Kalinova MG. Pollen storage at low temperature as a procedure for the improvement of cold tolerance in spring rape. Plant Breeding. 1998;**117**:389-391. DOI: 10.1111/j.1439-0523.1998.tb01959.x

[8] Lyakh V, Soroka A, Mishchenko L. Flowering time in oil flax can be influenced by microgametophytic selection. Euphytica. 2001;**118**(3):237-242. DOI: 10.1023/a:1017516532724

[9] Meghana KJ, Ravikumar RL. Effect of pollen selection for moisture stress tolerance in maize (*Zea mays* L.). Mysore Journal of Agricultural Sciences. 2018;**52**(2):340-344. Available from: https://www.cabdirect.org/cabdirect/abstract/20193384212

[10] Mishchenko LY. Microgametophytic Selection for Tolerance to Abiotic Stresses and Earliness in Oil Flax. Zaporozhye: Institute of Oilseed Crops UAAS; 1999 (in Ukrainian)

[11] Ottaviano E, Sari Gorla M, Mulcahy DL. Pollen selection: Efficiency and monitoring. In: Isozymes: Structure, Function and Use in Biology and medicine. New York: Wiley-Liss, Inc; 1990. pp. 575-588

[12] Rani TS, Ravikumar RL. Sporophytic and gametophytic recurrent selection for improvement of partial resistance to Alternaria leaf blight in sunflower (*Helianthus annuus* L.). Euphytica. 2006;**147**:421-431. DOI: 10.1007/s10681-005-9039-6

[13] Ravikumar RL, Patil BS, Salimath PM. Drought tolerance in sorghum by pollen selection using osmotic stress. Euphytica. 2003;**133**:371-376. DOI: 10.1023/A:1025702709095

[14] Winsor JA, Davis LE, Stephenson AG. The relationship between pollen load and fruit maturation and the effect of pollen load on offspring vigour in Cucurbita pepo. The American Naturalist. 1987;**129**(5):643-656. Available from: https://www.journals.uchicago.edu/doi/abs/10.1086/284664

[15] Mulcahy DL. Gametophytic gene expression. In: Blonstein AD, King PJ, editors. A Genetic Approach to Plant Biochemistry. Plant Gene Research (Basic Knowledge and Application). Vienna: Springer; 1986. pp. 247-258

[16] Ottaviano E, Mulcahy DL. Genetics of angiosperm pollen. Advances in Genetics. 1989;**26**:1-64

[17] Lyakh VA. Microgametophytic selection and its role in evolution of the angiosperms. Cytology and Genetics C/C of Tsitologiia I Genetika. 1995;**29**(6):76-82

[18] Rauf S. Breeding sunflower (helianthus annuusL.) for drought tolerance. Communications in Biometry and Crop Science. 2008;**3**:29-44. Available from: https://www.researchgate.net/publication/26514707_Breeding_sunflower_Helianthus_annuus_L_for_drought_tolerance

[19] Lyakh V, Soroka A, Vasin V. Influence of mature and immature sunflower seed treatment with ethylmethanesulphonate on mutation spectrum and frequency. Helia. 2005;**28**:87-98. DOI: 10.2298/HEL0543087L

[20] Soroka AI, Lyakh VA. Inheritance of two types of modified leaf venation in sunflower (Helianthus annuus L.). Indian Journal of Genetics and Plant Breeding. 2015;**75**(1):75-78. Available from: http://epubs.icar.org.in/ejournal/index.php/IJGPB/article/view/63638

[21] Patil BS, Ravikumar RL, Salimath PM. Effect of pollen selection for moisture stress tolerance on progeny performance in sorghum. Journal of Food, Agriculture and Environment. 2006;**4**:201-204

[22] Polevoy VV, Chirkova TV, editors. Practical Work on Plant Growth and Resistance: Study Guide. St. Petersburg: St.Petersburg University; 2001. p. 212 (in Russian)

[23] Totsky IV, Lyakh VA. Pollen selection for drought tolerance in sunflower. Helia. 2015;**38**(63):211-220. DOI: 10.1515/helia-2015-0012

[24] Lyakh V, Totsky I. Selective elimination of gametes during pollen storage at low temperature as a way to improve the genetic structure of sporophytic population for coldtolerance. Helia. 2014;**37**:227-235. DOI: 10.1515/helia-2014-0021

[25] Lyakh VA, Soroka AI, Mishchenko LY, Kalinova MG, Miroshnichenko EN. The Methods of Selection of Valuable Genotypes at the Level of Pollen. Zaporizhzhya: Institute of Oilseed Crops UAAS; 2000. p. 48 (in Russian)

[26] Lyakh V, Totsky I. Heat tolerance and adaptability to drought can be influenced by pollen selection. Helia. 2014;**37**:77-86. DOI: 10.1515/helia-2014-0006

[27] Lyakh VA, Gasenko NV, Soroka AI. Influence of pollen heating on the quality of resulting sporophyte generation in sunflower. Helia. 1998;**21**(29):103-108. Available from: https://www.researchgate.net/publication/275352224

[28] Tarakanov GI, Dovedar SA, Avakimova LG, Andreeva LN, Sysina EA. Methods of increasing fruit set in tomato under high temperature conditions. In: Genotype and Environment in Breeding Greenhouse Tomatoes (Materials of the EUCARPIA Meeting). Leningrad; 1978. pp. 123-129. Available from: https://eurekamag.com/research/000/698/000698330.php

[29] Domınguez E, Cuartero J, Fernandez-Munoz R. Breeding tomato for pollen tolerance to low temperatures by gametophytic selection. Euphytica. 2005;**142**:253-263. DOI: 10.1007/s10681-005-2042-0

Chapter 8

Impacts of Drought on Homestead Plant Diversity in Barind Tract of Bangladesh

Md. Shafiqul Islam and Md. Nazrul Islam Mukul

Abstract

Homestead is a great place for household food access, diet, and nutrition. Drought affects homestead plant diversity and reduces production, availability, and diversity that lead toward less supply and consumption. Drought detains moisture and degrades the soil that supports plant growth. Homestead provides regular bread and income in the rural areas with an effective means for both economic and environmental well-being. People are getting a good amount of subsidiary income without any extra care and effort. In managing homestead land and drought, the household needs necessary technical and managerial training. In reducing drought effects to the homestead, action research needs to be carried out on available knowledge, effective practices, water management, and the adoption of local varieties and knowledge to develop effective homestead integration. Government initiatives, community engagement and not harming the environment, and efficient uses of water could be great solutions for the adverse effects of drought on the homestead plant diversity.

Keywords: barind tract, biodiversity, drought, environment, homesteads, plant diversity

1. Introduction

The most complex and uncertain natural disaster is drought, even it is difficult to predict and mitigate due to numerous factors, magnitude as well as difficulties in defining a drought. Drought causes impacts on agro biodiversity in many ways [1]. Agro biodiversity is the part of biodiversity includes both genetic diversity and crop and animal diversity in diverse agro ecosystem and different agriculture [2, 3]. It was mentioned that agro biodiversity is essentially important for sustainable development both for natural and anthropogenic as well as livelihood [4, 5]. Homestead is the traditional place and promising option for plant species conservation [6]. These efforts may help mitigate ecosystem degradation and drought impacts while providing food and economic opportunities to rural people. In the homestead, members share responsibilities and distribute gardening tasks among themselves. Women like to use homestead products for household consumption and gifting and men for sale and cash income [7]. Numerous studies in the world showed that homesteads are the place of the highest plant diversity for crops and non-crops [8]. A study mentioned that agro biodiversity support sustainable agriculture [9]. Food and Agricultural Organization-FAO [10, 11] estimated that there are 2, 50,000 to 5, 00,000 plant species in the world. According to a study, in agriculture only 1,500 plant species are being used. Another study mentioned that agrobiodiversity exists at the varietal level though number of crop species is low [12]. Agrobiodiversity provide three types of value such as use value, option value, and existence value [13]. Ecosystems that maintain naturally are considered as the major support for sustainable production and enhance food and livelihood security at local to global scale [2, 5, 14]. Agro biodiversity is contributing to agricultural production, maintaining ecosystems and human food security [12, 15–17]. Ecosystem provides four major categories of services include provisioning services, supporting services, regulating services and cultural services. Benefit may get through intrinsic values (cultural and social) and supported by biological diverse system [15].

Agro biodiversity provides diverse benefits including nutritious diets (fruits and vegetables) for human health; longer productivity; adaptation to changing conditions; and conservation methods for enabling future use. Biological diversity reduces the general risk of economic and ecological failure at individual (farmer) level as well as on global scale while making those systems more sustainable and less susceptible facing the future challenges [17–19]. Diverse agricultural production system may be less threatening to biological diversity compared to highly intensified and uniform systems, such practices still do enhance pest damage, competition for water, soil, nutrition and pollinators, result in habitat and nutrient loss and have a negative effect on non-target species [20]. The reduction of agro biodiversity is closely linked to the loss of productivity, while threatening ecological stability, the security of food supply and livelihood worldwide [15]. Strategies for identification and conservation of agro biodiversity are lacking and hence, there is need to develop indices for identification of important component of agro biodiversity, food and nutritional security. Agro biodiversity has been disturbed due to switching from rain-fed agriculture to irrigated agriculture. Increasing frequency of extreme weather patterns and droughts are going to be the most important climatic factors affecting agro biodiversity [21, 22]. Diversity has been accounted at three levels such as alpha, gamma and beta [23]. Diversity at alpha level could be measured at community level, beta for composition (change from farmers to farmers) and gamma at the larger units [2, 24]. The assessment of drought impacts on homestead plant diversity in the Barind Tract is essentially important. The study was conducted with the view to assess and compare drought impacts on homestead plant diversity in high drought and low drought areas of Bangladesh. The study is also focused on following objectives:

- Find out comparative plant diversity between drought and low drought prone areas at the homestead level in relation to season

- Establish the relationship with the homestead plant diversity with the variables

- Estimate plant species (Timber, spices, fruits, ornamental and vegetables) diversity in the homestead

The research elicited following research questions:
Research Questions:

- How they perceived drought and plant diversity?

- Are there any significant changes in homestead plant biodiversity between drought and non-drought prone areas?

- Is there any relationship between drought and plant biodiversity?
- How drought impacted on species (plants) at the homesteads?

Research Hypotheses:
H_1: There is a trend of greater homestead plant diversity in the low drought prone area than high drought prone area

2. Methodology

The study was conducted over the period of twelve months starting from July 2018 and ended by May 2019. Two sites were selected for this study one from high drought prone site and another from low drought prone site of Barind Tract with the view to see the impacts of drought on the homestead plant diversity. Homestead agro biodiversity (plants) was calculated using Shannon's Index. Homestead agro biodiversity (only plant species) was measured using similar size of homestead from different group (small, medium and large) as the quadrant. Plant species was studied using quadrat method covering random quadrat of 10 m × 10 m size was laid for trees, vegetables and other plants (shrubs and herbs) were laid in each homestead. Total 300 respondents (150 respondents from each site) considering equal numbers of respondents from small, medium and large category of the homestead. The farm owners were asked to know about the plant species and dietary diversity. Eight Focus Group Discussion (four from each site: two with elderly people and two with the ethnic people) have been carried out to know the insights of plant diversity at the homesteads. Semi-structured questionnaire, guided checklist, expert consultation and field observation has been carried out for this study. Quadrant method has been followed in collecting data from sample unit in the study areas. Nine quadrates from high drought prone areas and nine quadrates from low drought prone areas (3 from each farm size) have been laid for the measurement agro biodiversity. Survey on crops (species) has been carried out in the sampled quadrant. Secondary relevant information has been used to enrich the study. The agro biodiversity has been calculated using Shannon's diversity Index.

$$H = -\sum_{i=1}^{S} p_i \ln p_i$$

H = Shannon's diversity index.
S = Total number of species in the community (richness).
p_i = The proportion of species i relative to the total number of species (p_i).
$\ln p_i$ = The natural logarithm of this proportion.

3. Results

3.1 Perception of drought impacts on homestead plant diversity

An assessment on perception of drought impacts to the homestead plant diversity was carried out using Rating Percentage Index (RPI). The higher rate of respondents (RPI 30) from high drought areas perceived that plant diversity is decreasing in the homestead due to drought and low rate of respondents (RPI 42.5) from low drought areas assumed that homestead plant diversity is decreasing due to drought (**Table 1**). The participants from Focus Group Discussions opined that low diversity reduces production, consumption, dietary diversity and household nutrition.

3.2 Homestead plant species diversity in different seasons

3.2.1 Homestead agro biodiversity in the Rabi season

There is a negative impact of drought in the study areas. Results showed that low species diversity in the high drought prone areas during Rabi season (**Table 2**). The low diversity is the spiraling issues of drought. The occurrence of rainfall in the high drought prone areas is low and erratic that give birth low soil moisture resulting low yields or harvests. Farmers are trying to grow several crops for better yield and managing drought. Total 38 plant species were recorded in the homesteads from high drought prone areas and 34 plant species in the low drought homestead areas. Several causes such as water and moisture scarcity, land degradation, soil infertility, lack of awareness, less rainfall, high temperature and climate change issues have been identified by the respondents for low species diversity in the homesteads.

3.2.2 Homestead agro biodiversity in the Kharif-1 season

Results showed that the plant species diversity is more in low drought homesteads and low in the high drought homesteads (**Table 3**).

3.2.3 Homestead plant species diversity in the Kharif-2 season

It was found that homesteads of low drought areas show greater diversity than high drought area (1.287 ‹ 7.294) during Khari-2 season (**Table 4**). Respondents mentioned that it happened due to low rainfall and low soil moisture retention during the period.

3.3 Relationship of plant species diversity with the seasons and variables

The respondents believe that it is happening due to drought (causal effects of low soil moisture, low water holding capacity of the soil, low soil fertility and low rainfall over the season). It also affects agro ecology for the production. The numbers of family member has the effect on homestead agro biodiversity followed by age,

Homestead diversity	High drought prone areas			Low drought prone areas		
	Perception of acceptance	Perception of rejection	RPI	Perception on acceptance	Perception on rejection	RPI
Decreasing	80	20	30	30	70	42.5
Increasing	0	100	50	10	90	47.5

Table 1.
Perception of drought impacts on homestead plant diversity (field data, 2019).

DOI: http://dx.doi.org/10.5772/intechopen.101885

Name of Species	High Drought Prone area				Low Drought Prone area				
	Total	PI	lnpi	-(−pi*lnpi)	Name of Species	Total	PI	lnpi	-(−pi*lnpi)
Mangifera indica	154	0.25581	−0.59208	0.151	*M. indica*	94	0.2117	−0.6743	0.14275
Artocarpus heterophyllus	9	0.01495	−1.82535	0.027	*A. heterophyllus*	8	0.018	−1.7443	0.03143
Psidium guava	17	0.02824	−1.54915	0.044	*Psidium guava*	19	0.0428	−1.3686	0.05857
Lagenaria siceraria	16	0.02658	−1.57548	0.042	*L. siceraria*	22	0.0495	−1.305	0.06466
Azadirachta indica	46	0.07641	−1.11684	0.085	*Spondias mombin*	7	0.0158	−1.8023	0.02841
S. mombin	12	0.01993	−1.70042	0.034	*Ziziphus mauritiana*	15	0.0338	−1.4713	0.04971
Z. mauritiana	14	0.02326	−1.63347	0.038	*Moringa oleifera*	11	0.0248	−1.606	0.03979
M. oleifera	10	0.01661	−1.7796	0.03	*Cucurbita*	9	0.0203	−1.6931	0.03432
Aegle marmelos	8	0.01329	−1.87651	0.025	*Citrus Limonium*	14	0.0315	−1.5013	0.04734
Amorphophallus paeoniifolius	7	0.01163	−1.9345	0.022	*Musa paradisicum*	19	0.0428	−1.3686	0.05857
Colocasia esculenta	41	0.06811	−1.16681	0.079	*Annona reticulate*	4	0.009	−2.0453	0.01843
Bambusa spp	4	0.00664	−2.17754	0.014	*Tamarindus indica*	2	0.0045	−2.3464	0.01057
Citrus limonium	68	0.11296	−0.94709	0.107	*Rubus*	7	0.0158	−1.8023	0.02841
Musa paradisicum	14	0.02326	−1.63347	0.038	*Litchi chinensis*	4	0.009	−2.0453	0.01843
Annona reticulate	67	0.1113	−0.95352	0.106	*Citrus maxima*	4	0.009	−2.0453	0.01843
Saccharum officinarum	2	0.00332	−2.47857	0.008	*Syzygium jambos*	3	0.0068	−2.1703	0.01466
Punica granatum	25	0.04153	−1.38166	0.057	*Swietenia mahagoni*	26	0.0586	−1.2324	0.07217
Momordica charantia	3	0.00498	−2.30248	0.011	*Tagetes*	34	0.0766	−1.1159	0.08545
Capsicum frutescens	2	0.00332	−2.47857	0.008	*Phaseolus vulgaris*	13	0.0293	−1.5334	0.0449
Dioscorea	15	0.02492	−1.60351	0.04	*Areca palm*	7	0.0158	−1.8023	0.02841
C. frutescens	7	0.01163	−1.9345	0.022	*Rosa sinensis*	19	0.0428	−1.3686	0.05857
Jiga	5	0.00831	−2.08063	0.017	*Carissa carandas*	2	0.0045	−2.3464	0.01057

Name of Species	High Drought Prone area				Low Drought Prone area				
	Total	PI	lnpi	-(–pi*lnpi)	Name of Species	Total	PI	lnpi	-(–pi*lnpi)
T. indica	2	0.00332	–2.47857	0.008	*Syzygium aqueum*	6	0.0135	–1.8692	0.02526
Rubus	3	0.00498	–2.30248	0.011	*Carica papaya*	16	0.036	–1.4433	0.05201
Phoenix dactylifera	13	0.02159	–1.66565	0.036	*Basella alba*	14	0.0315	–1.5013	0.04734
L. chinensis	1	0.00166	–2.7796	0.005	*Ocimum tenuifloru*	7	0.0158	–1.8023	0.02841
Citrus × paradise	1	0.00166	–2.7796	0.005	*Jasminum*	5	0.0113	–1.9484	0.02194
S. jambos	5	0.00831	–2.08063	0.017	*Cocos nucifera*	26	0.0586	–1.2324	0.07217
S. mahagoni	12	0.01993	–1.70042	0.034	*Dillenia indica*	1	0.0023	–2.6474	0.00596
Tagetes	3	0.00498	–2.30248	0.011	*Cestrum nocturnum*	1	0.0023	–2.6474	0.00596
P. vulgaris	2	0.00332	–2.47857	0.008	*Ficus racemosa*	2	0.0045	–2.3464	0.01057
Arecacpalm	3	0.00498	–2.30248	0.011	*Manilkara zapota*	2	0.0045	–2.3464	0.01057
Rosa sinensis	3	0.00498	–2.30248	0.011	Ghasful	8	0.018	–1.7443	0.03143
Dalbergia sissoo	2	0.00332	–2.47857	0.008	*Cascabela thevetia*	2	0.0045	–2.3464	0.01057
C. carandas	2	0.00332	–2.47857	0.008		444	0.9752		1.28672
S. mombin	2	0.00332	–2.47857	0.008					
C. papaya	1	0.00166	–2.7796	0.005					
Bombax cieba	1	0.00166	–2.7796	0.005					
	602	1		1.201					
Shanon Index (H)				1.201	Shanon Index (H)				1.286
Diversity (D)				9.239	Diversity (D)				12.7659

Table 2.
Plant species diversity in the homestead in Rabi season (field data, 2019).

DOI: http://dx.doi.org/10.5772/intechopen.101885

High drought prone area					**Low drought prone area**				
Species	**Ni**	**Pi**	**ln Pi**	**-(Pi*ln Pi)**	**Species**	**Ni**	**Pi**	**ln Pi**	**-(Pi*ln Pi)**
Mangifera indica	171	0.299	−0.524	0.157	*M. indica*	104	0.219	−0.659	0.145
Artocarpus heterophyllus	13	0.023	−1.643	0.037	*A. heterophyllus*	8	0.017	−1.773	0.030
Psidium guava	17	0.030	−1.526	0.045	*Psidium guava*	19	0.040	−1.397	0.056
Lagenaria siceraria	12	0.021	−1.677	0.035	*L. siceraria*	22	0.046	−1.333	0.062
Azadirachta indica	52	0.091	−1.041	0.095	*Spondias mombin*	7	0.015	−1.831	0.027
S. mombin	12	0.021	−1.677	0.035	*Ziziphus*	15	0.032	−1.500	0.047
Ziziphus	14	0.025	−1.611	0.039	*Moringa oleifera*	11	0.023	−1.634	0.038
M. oleifera	10	0.018	−1.757	0.031	Bel*(Aegle marmelos)*	2	0.004	−2.375	0.010
Bel*(A. marmelos)*	8	0.014	−1.854	0.026	*(Bambusoideae*	46	0.097	−1.013	0.098
Amorphophallus paeoniifolius	41	0.072	−1.144	0.082	*Citrus limonium*	14	0.030	−1.530	0.045
Bamboo*(Bambusoideae)*	68	0.119	−0.924	0.110	*Musa paradisicum*	19	0.040	−1.397	0.056
Lemon*(Citrus Limonium)*	14	0.025	−1.611	0.039	*amarindus indica*	2	0.004	−2.375	0.010
Musa paradisicum	67	0.117	−0.931	0.109	Blackberry*(Rubus)*	7	0.015	−1.831	0.027
Punica granatum	3	0.005	−2.280	0.012	*Phoenix dactylifera*	5	0.011	−1.977	0.021
Momordica charantia	1	0.002	−2.757	0.005	Litchi*(Litchi chinensis)*	4	0.008	−2.074	0.017
Yam*(Dioscorea)*	15	0.026	−1.581	0.042	*Citrus maxima*	4	0.008	−2.074	0.017
Capsicum frutescen	7	0.012	−1.912	0.023	*Syzygium jambos*	3	0.006	−2.199	0.014
Luffa acutangula	5	0.009	−2.058	0.018	*Swietenia species*	26	0.055	−1.261	0.069
Tamarindus indica	2	0.004	−2.456	0.009	*Tagetes*	34	0.072	−1.144	0.082
Blackberry*(Rubus)*	3	0.005	−2.280	0.012	Palm*(Arecaceae)*	7	0.015	−1.831	0.027
P. dactylifera	13	0.023	−1.643	0.037	Rose*(Rosa)*	19	0.040	−1.397	0.056
Litchi*(L. chinensis)*	1	0.002	−2.757	0.005	*Carissa carandas*	2	0.004	−2.375	0.010

High drought prone area					Low drought prone area				
Species	**Ni**	**Pi**	**In Pi**	**-(Pi*ln Pi)**	**Species**	**Ni**	**Pi**	**ln Pi**	**-(Pi*ln Pi)**
Mahagoni(*Swietenia*)	12	0.021	−1.677	0.035	*Carica papaya*	16	0.034	−1.472	−0.050
Marigold(*Tagetes*)	3	0.005	−2.280	0.012	Ccurbita	9	0.019	−1.722	0.033
Palm(*Arecaceae*)	3	0.005	−2.280	0.012	Annona reticulate	4	0.008	−2.074	0.017
Dalbergia sissoo	2	0.004	−2.456	0.009	*Syzygium aqueum*	6	0.013	−1.898	0.024
C. papaya	1	0.002	−2.757	0.005	Basellla alba	14	0.030	−1.530	0.045
Bombax cieba	1	0.002	−2.757	0.005	*Ocimum sanctum*	7	0.015	−1.831	0.027
					Jasminum	5	0.011	−1.977	0.021
					Cocos nucifera	26	0.055	−1.261	0.069
					Chalta(*Dillenia indica*)	1	0.002	−2.676	0.006
					Ficus racemosa	2	0.004	−2.375	0.010
					Manilkara zapota	2	0.004	−2.375	0.010
					Cascabela thevetia	2	0.004	−2.375	0.010
Total	571				Total	474			1.287
Shanon Index (H)				1.082	Shanon Index (H)				1.287
Diversity (D)				7.294	Diversity (D)				12.14

Table 3.
Plant species diversity in the homestead in Kharif-1 season (field data, 2019).

High drought prone area					Low drought prone area				
Species	Ni	Pi	ln Pi	-(Pi*ln Pi)	Species	Ni	Pi	ln Pi	-(Pi*ln Pi)
Mangifera indica	125	0.22	−0.65	0.145028	*M. indica*	67	0.218	−0.66	0.14
Artocarpus heterophyllus	9	0.02	−1.80	0.028676	*A. heterophyllus*	7	0.023	−1.64	0.04
Psidium guava	15	0.03	−1.57	0.041893	*Psidium guava*	16	0.052	−1.28	0.07
Lagenaria siceraria	16	0.03	−1.54	0.043891	*L. siceraria*	15	0.049	−1.31	0.07
Azadirachta indica	45	0.08	−1.09	0.088780	*Spondias mombin*	5	0.016	−1.79	0.03
S. mombin	12	0.02	−1.67	0.035576	*Ziziphus*	8	0.026	−1.59	0.04
Ziziphus	13	0.02	−1.64	0.037740	*Moringa oleifera*	7	0.023	−1.64	0.04
M. oleifera	10	0.02	−1.75	0.031051	sweet gourd(*Cucurbita*)	6	0.019	−1.71	0.03
Bel(*Aegle marmelos*)	8	0.01	−1.85	0.026215	*Citrus Limonium*	9	0.029	−1.53	0.04
sweet gourd(*Cucurbita*)	6	0.01	−1.97	0.020990	*Musa paradisicum*	7	0.023	−1.64	0.04
Amorphophallus paeoniifolius	47	0.08	−1.08	0.089931	Ata(*Annona reticulate*)	3	0.01	−2.01	0.02
Colocasia esculenta	5	0.01	−2.05	0.018194	*Tamarindus indica*	2	0.006	−2.19	0.01
Bamboo(*Bambusoideae*)	65	0.12	−0.94	0.108144	Blackberry(*Rubus*)	7	0.023	−1.64	0.04
Lemon(*Citrus Limonium*)	14	0.02	−1.61	0.039844	Litchi(*Litchi chinensis*)	4	0.013	−1.89	0.025
Banana(*Musa paradisicum*)	59	0.10	−0.98	0.102562	Jambura(*Citrus maxima*)	4	0.013	−1.89	0.02
Ata(*Annona reticulate*)	2	0.01	−2.45	0.008688	*Syzygium jambos*	1	0.003	−2.49	0.01
Saccharum officinarum	25	0.04	−1.35	0.059988	Mahagoni(*Swietenia*)	14	0.045	−1.34	0.06
Bedena (*Punica granatum*)	3	0.01	−2.27	0.012096	Marigold(*Tagetes*)	29	0.094	−1.03	0.10
Momordica charantia	2	0.03	−2.45	0.008688	*Phaseolus vulgaris*	9	0.029	−1.53	0.04
Yam(*Dioscorea*)	15	0.03	−1.58	0.041893	Palm(*Arecaceae*)	5	0.016	−1.79	0.03
Chili (*Capsicum frutescens*)	7	0.01	−1.91	0.023658	Rose(*Rosa*)	11	0.036	−1.45	0.05
Jhinga (*Luffa acutangula*)	5	0.01	−2.05	0.018194	*Carissa carandas*	1	0.003	−2.49	0.01

High drought prone area					Low drought prone area				
Species	**Ni**	**Pi**	**In Pi**	**-(Pi*ln Pi)**	**Species**	**Ni**	**Pi**	**ln Pi**	**-(Pi*ln Pi)**
Tamarind(*T. indica*)	2	0.00	−2.45	0.008688	*(Syzygium aqueum)*	6	0.019	−1.71	0.03
Blackberry(*Rubus*)	3	0.01	−2.27	0.012096	Papaya(*Carica papaya*)	13	0.042	−1.37	0.06
Phoenix dactylifera	14	0.02	−1.61	0.039844	*Basella alba*	7	0.023	−1.64	0.04
Litchi(*L. chinensis*)	2	0.00	−2.45	0.008688	*Ocimum tenuifloru*	7	0.023	−1.64	0.04
Jambura(*C. maxima*)	1	0.00	−2.75	0.004878	*Jasminum*	4	0.013	−1.89	0.02
Golap jam(*S. jambos*)	5	0.01	−2.05	0.018194	Coconut(*Cocos nucifera*)	22	0.071	−1.15	0.09
Mahagoni(*Swietenia*)	13	0.02	−1.64	0.037740	Chalta(*Dillenia indica*)	1	0.003	−2.49	0.01
Marigold(*Tagetes*)	3	0.01	−2.27	0.012096	*Cestrum nocturnum*	1	0.003	−2.49	0.01
P. vulgaris	2	0.00	−2.45	0.008688	*Ficus racemosa*	2	0.006	−2.19	0.01
Palm(*Arecaceae*)	4	0.01	−2.15	0.015242	*Manilkara zapota*	2	0.006	−2.19	0.01
Rose(*Rosa*)	2	0.00	−2.45	0.008688	Ghasful	5	0.016	−1.79	0.03
Sissoo(*Dalbergia sissoo*)	2	0.00	−2.45	0.008688	*Cascabela thevetia*	1	0.003	−2.49	0.01
C. carandas	2	0.00	−2.45	0.008688	Total	308	1	−59.58	1.31
Shanon Index (H)				1.082	Shanon Index (H)				1.287
Diversity (D)				7.294	Diversity (D)				12.14

Table 4.
Homestead diversity inKhari-2 (field data, 2019).

income, education, land size and farm category. According to the hypothesis there is significant difference of homestead agro biodiversity between high and low drought prone areas. The family member has the large effect on homestead agro biodiversity followed by age, income, education and farm category (**Table 5**).

3.4 Homestead plant species diversity changing over times

As per Focus Group Discussions, the participants mentioned that plant diversity is decreasing in the high drought prone areas due to drought effects. Seventy seven percent household respondents from high drought prone areas reported that homestead plant diversity is decreasing with the increased frequency of drought (**Figure 1**). Only 10 percent household respondents from low drought prone areas believe that homestead plant diversity is decreasing with the influence of drought.

3.5 Drought impacts on the homestead production

It was found that 30 percent respondents perceived reduced plant species diversity in the high drought area. People in the low drought area are confused about the changing of agro biodiversity. About 53% people of low drought area observed extinction of some species (**Figure 2**).

Diversity value	Age	Income	Education	Farm category	Tenanat type	Location	member
0.134199134	38	140000	9	1	1	1	3
0.190675991	60	120000	0	1	3	1	6
0.131275168	46	72000	8	1	1	1	4
0.092691622	45	200000	1	2	1	1	4
0.087486157	42	70000	5	2	1	1	5
0.126855601	42	300000	8	2	1	1	6
0.089473684	46	300000	0	3	1	1	5
0.177253479	42	325000	10	3	3	1	5
0.260814249	60	310000	10	3	1	1	6
0.060846561	60	120000	10	1	3	2	2
0.111111111	33	35000	1	1	1	2	2
0.169117647	38	100000	1	1	3	2	5
0.070075758	53	100000	10	2	3	2	3
0.06685633	37	150000	8	2	3	2	4
0.083870968	52	175000	5	2	1	2	4
0.046444122	52	280000	0	3	3	2	4
0.109311741	50	245000	10	3	1	2	4
0.180225989	65	80000	8	3	3	2	4
r	0.25	0.15	0.16	0.08	−0.02	−0.40	0.56

Table 5.
Correlation of homestead plant diversity among the variables.

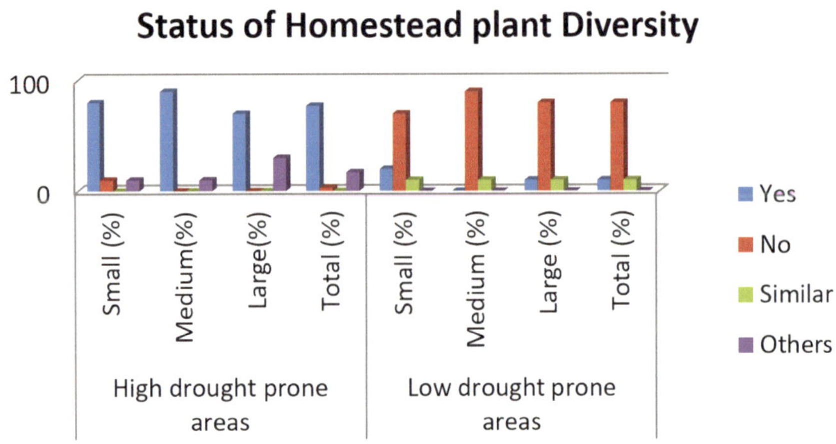

Figure 1.
status of homestead plant diversity in relation to farm category.

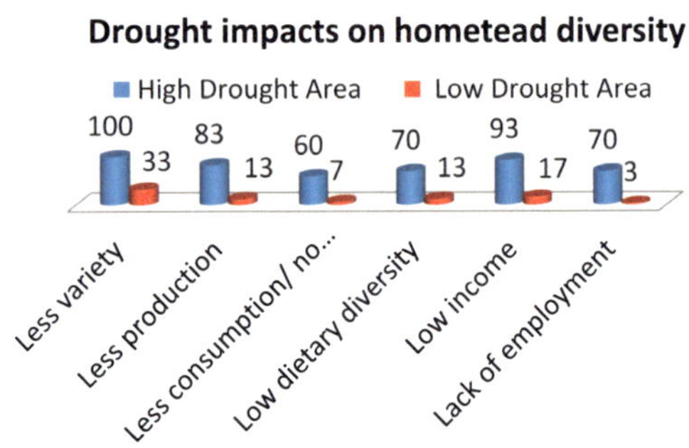

Figure 2.
Drought impacts on homestead diversity.

Homestead number	High drought prone area	Low drought prone area
H1	0.134199	0.060847
H2	0.190676	0.111111
H3	0.131275	0.169118
H4	0.092692	0.070076
H5	0.087486	0.066856
H6	0.126856	0.083871
H7	0.089474	0.046444
H8	0.177253	0.109312
H9	0.260814	0.180226

Table 6.
Species diversity (Ds) value of high drought and low drought prone area.

3.6 Testing of hypothesis

Low drought prone area tends to a greater homestead agro bio-diversity than high drought prone area. Calculated t value 0.004555 and table value with the df (2–1)(9–1) = 8 at 99% level is 2.896, i.e. the calculated t value is less than critical t value

(0.004555 ‹ 2.896). The hypothesis is accepted and there is significant greater homestead agro biodiversity in low drought prone area (**Table 6**).

4. Discussion

It was found that homestead number five (medium farm) showed greater plant diversity in the high drought prone areas and homestead number seven (large farm) showed greater diversity in the low drought prone areas. A study reported that the small farm holders utilized their homesteads more efficiently [25]. The study found that the homesteads contained tree species, fruit tree species, vegetable species, spices, ornamental species, medicinal species and wild species for different uses (timber, fuel, food, medicine and esthetic). Total 43 plant species (trees, fruit, vegetables, spices, ornamental, fuel and medicinal) were found in the study areas. Similarly 43 plant species were reported in the homesteads from the central parts of Bangladesh [26]. Timber tree got more preferences followed by fruits, vegetables, spices and ornamental. The researcher reported that timber tree species have been found as the preferred species due to future capital as the wood [25]. The results showed that the homestead in the low drought prone areas represent greater plant diversity in the homesteads. Household use their homestead produces mostly for foods, fuel, medicinal purpose and cash income. One study concluded that home gardens can be a tool for conservation of biodiversity [27]. Homestead plant diversity is essentially important to cope with the climate change issues and drought. Similarly a study reported that homestead biodiversity enhance the resilience capacity of the community [28]. Another study found that diverse variety increases production especially in environmental extreme (drought, flood) situations, they maintain diversity to reduce the risks [29]. In our existence, more plant species biodiversity can secure food for the farming community. A study is also reported that it is hard in surviving severe drought frequently cause famine in farming communities [30].

5. Conclusion

Homestead is the great place for household food access, diet and nutrition. Drought affects homestead plant diversity and reduces production, availability and diversity that lead toward less supply and consumption. Drought arrest moisture and degrade the soil that support plant growth. Actually homestead provides regular bread and income in the rural areas with an effective mean for both economic and environmental well-being. People are getting good amount of subsidiary income without any extra care and effort. In managing homestead land and drought, household needs necessary technical and managerial training. In reducing drought impacts to the homestead, action research needs to be carried out on available knowledge, effective practices, water management and adoption of local varieties and knowledge in order to develop effective homestead integration. Government initiatives, community engagement and doing no harm to the environment and efficient uses of water could be great solutions for adverse effects of drought to the homestead plant diversity.

Author details

Md. Shafiqul Islam[1*] and Md. Nazrul Islam Mukul[2]

1 State University of Bangladesh, Dhaka, Bangladesh

2 Cascade Architects Ltd, Dhaka, Bangladesh

Address all correspondence to: ms.du2019@gmail.com

References

[1] Belle JA, Hlalele MB. Vulnerability assessment of agricultural drought hazard: A case of koti-se-phola community council, Thabana Morena, Mafeteng district in Lesotho. J Geogr Nat Disast. 2015;**5**(143):2167-0587

[2] Love B, Spaner D. Agrobiodiversity: Its value, measurement, and conservation in the context of sustainable agriculture. Journal of Sustainable Agriculture. 2007;**31**(2):53-82

[3] Vandermeer J, Perfecto I. Breakfast of Biodiversity: The Political Ecology of Rain Forest Destruction. Food First Books; 2013

[4] Frison EA, Cherfas J, Hodgkin T. Agricultural biodiversity is essential for a sustainable improvement in food and nutrition security. Sustainability. 2011;**3**:238-253

[5] Thrupp LA. Linking agricultural biodiversity and food security: The valuable role of agrobiodiversity for sustainable agriculture. International Affairs. 2000;**76**(2):283-297

[6] Gardner TA, Barlow J, Chazdon R, Ewers RM, Harvey CA, Peres CA, et al. Prospects for tropical forest biodiversity in a human-modified world. Ecol Lett. 2009;**12**:561-582. DOI: 10.1111/j.1461-0248.2009.01294.x

[7] Reyes-García V, Vila S, Aceituno-Mata L, Calvet-Mir L, Garnatje T, Jesch A, et al. Gendered homegardens: A study in three mountain areas of the Iberian Peninsula. Econ Bot. 2010;**64**:235-247. DOI: 10.1007/s12231-010-9124-1

[8] Coomes OT, Ban N. Cultivated plant species diversity in home gardens of an Amazonian peasant village in northeastern Peru. Econ Bot. 2004;**58**:420-434

[9] Cleaveland DA, Soleri D, Smith SE. Do folk varieties have a role in sustainable agriculture? Incorporating folk varieties into the development of locally based agriculture may be the best approach. Bioscience. 1994;**44**:740-751

[10] FAO. The state of the world's plant genetic resources for food and agriculture. FAO; 1998

[11] Wilkes HG. Gerplasm collections: their use, potential, social responsibility, and genetic vulnerability. In: Buxton DR et al., editors. International Crop Science I. Madison, WI: Crop Science Society of America; 1993. pp. 445-450

[12] Brush SB. Farmers' Bounty: The Survival of Crop Diversity in the Modern World. United States: Yale University Press; 2004

[13] Pardey PG, Skovmand B, Taba S, Van Dusen ME, Wright BD. The cost of conserving maize and wheat genetic resources ex situ. In: Farmers Gene Banks and Crop Breeding: Economic Analyses of Diversity in Wheat Maize and Rice. Dordrecht: Springer; 1998. pp. 35-56

[14] FAO. Cultivating our future. sustaining the multiple functions of agricultural bio-diversity. Background Paper 1: Agricultural biodiversity. Maastricht, The Netherlands: Tech. Rep., Food and Agricultural Organization of the United Nation (FAO); 1999

[15] Cardinale BJ, Duffy JE, Gonzalez A, Hooper DU, Perrings C, Venail P, et al. Biodiversity loss and its impact on humanity. Nature. 2012;**486**(7401):59

[16] Esquinas-Alcázar J. Protecting crop genetic diversity for food security: Political, ethical and technical challenges. Nature Reviews Genetics. 2005;**6**(12):946

[17] Swift MJ, Izac AM, van Noordwijk M. Biodiversity and ecosystem services in agricultural landscapes—are we asking the right questions? Agriculture, Ecosystems & Environment. 2004; **104**(1):113-134

[18] Baumgärtner S, Quaas MF. Managing increasing environmental risks through agro biodiversity and agrienvironmentalpolicies. Agricultural Economics. 2010;**41**(5):483-496

[19] Di Falco S, Perrings C. Crop genetic diversity, productivity and stability of agroecosystems. A theoretical and empirical investigation. Scottish Journal of Political Economy. 2003;**50**(2):207-216

[20] Zhang W, Ricketts TH, Kremen C, Carney K, Swinton SM. Ecosystem services and dis-services to agriculture. Ecological Economics. 2007;**64**(2):253-260

[21] Kotschi J. Agricultural biodiversity is essential for adapting to climate change. GAIA-Ecological Perspectives for Science and Society. 2007;**16**(2):98-101

[22] Tisdell C. Socioeconomic causes of loss of animal genetic diversity: Analysis and assessment. Ecological Economics. 2003;**45**(3):365-376

[23] Whittaker RH. Dominance and diversity in land plant communities: numerical relations of species express the importance of competition in community function and evolution. Science. 1965;**147**(3655):250-260

[24] McCune B, Grace JB, Urban DL. Analysis of Ecological Communities. Vol. 28. Gleneden Beach, OR: MjM Software Design; 2002

[25] Baul TK, Rahman MM, Moniruzzaman M, Nandi R. Status, utilization, and conservation of agrobiodiversity in farms: a case study in the northwestern region of Bangladesh. International Journal of Biodiversity Science, Ecosystem Services & Management. 2015;**11**(4):318-329

[26] Muhammed N, Masum MFH, Hossain MM, Chakma S, Oesten G, von Detten R. Floral composition and biodiversity conservation in homestead forests in Mymensingh, Bangladesh. International Journal of Biodiversity Science, Ecosystem Services & Management. 2011;**7**(4):247-257

[27] Saikia P, Khan ML. Homegardens of upper Assam, northeast India: a typical example of on farm conservation of Agarwood (*Aquilaria malaccensis* Lam.). International Journal of Biodiversity Science, Ecosystem Services & Management. 2014;**10**(4):262-269

[28] Sundaray JK, Bhattacharya A, Ponniah AG, Ghoshal TK, Deo AD, Sharma JP, et al. Homestead farming: A biodiverse system to enhance resilience to climate vulnerability. In: Revitalizing the Ganges Coastal Zone: Turning Science into Policy and Practices Conference Proceedings. Colombo, Sri Lanka: CGIAR Challenge Program on Water and Food (CPWF). 600pp; 2015. p. 231

[29] Reda A, Tewelde F. Impact of invasive alien plants on biodiversity and livelihood of local communities in Tigray, Ethiopia. World Applied Sciences Journal. 2017;**35**(11):2364-2378

[30] Sircar S. (2019). Commercial shrimp farming in South-West Bangladesh. pub.epsilon.slu.se

Section 4

Water Risk

Chapter 9

Water Scarcity Management in the Maghreb Region

Kaltoum Belhassan

Abstract

The Maghreb region includes five countries Mauritania, Morocco, Algeria, Tunisia, and Libya. It is region of Northwest Africa bordering the Mediterranean Sea. The Maghreb consists of two defining regions, the Sahara Desert in the south and the Atlas Mountains in the north. The Maghreb has arid to semiarid climate. Many areas in the world and particularly the Maghreb countries are affected by water scarcity, as well as result of population pressures, rising in urbanization, climate change and also increasing pollution of water. This paper attempts to assess the region's water scarcity challenges through highlighting the causes and reasons of water scarcity and its negative effects on water supply. Also, the chapter aims to discuss the vital role of sustainable water management to reduce the risk of water scarcity and this through the solutions, techniques and the best practices adopted in the region.

Keywords: Maghreb, water scarcity, causes, management, solutions

1. Introduction

Although more than 71% of Earth is covered in water, 97.5% of the earth's water is salt water and most of the remaining 2.5% is locked away as groundwater or in glaciers; yet with such a vast reserve of water body on Earth, humans depend on the tiny bit available as fresh water and there is a global shortage of water. According to the WHO, it is estimated that by 2025, more than half of the world population could be living in areas facing water scarcity due to formation of mega-cities and increasing world population, which is expected to reach ~9.7B by 2050, causing further stress on water globally. Although water scarcity is a universal phenomenon, this chapter is focused on the continent of Africa and more specifically the region of Northwest Africa also known as the Maghreb. The region includes Algeria, Libya, Mauritania, Morocco and Tunisia (**Figure 1**). The Maghreb countries are located in an arid to semiarid region. The climate is very diverse and varying with the season and region; generally, it is characterized by mild-wet winters and warm-dry summers. Also, the climate in the Maghreb is Saharian in the South part, oceanic in the western, and Mediterranean in the North part. Although, each country in Northwest Africa has its own individual water reserves, regional characteristics and water management history, in spite of Algeria, Morocco and Tunisia have several similarities. A lot of data exists on the models and projections of water scarcity and as per Water Resource

Figure 1.
Map of the Maghreb-Northwest Africa [1].

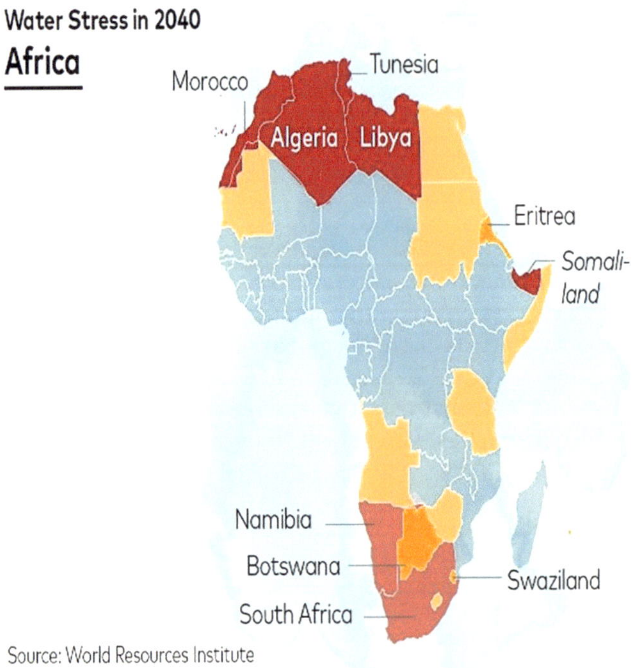

Figure 2.
As comparison projected water stress in Africa in 2040, as per water resource institute [2].

Institute, a map of water scarcity is shown in **Figure 2**, with areas in red showing extreme stress.

The Maghreb is one of the most "water-stressed" regions because of combinations of several factors, viz. population growth, climate change and anthropogenic contamination (anthropogenic pollution on the ecosystem). Tunisia, Algeria, Morocco and Libya face demand for water that is higher than the amount of quality water available-per capita; water availability has fallen below 1000 m^3 annually. These countries are all on the top 33 water-stressed countries in the world [3]. However, Mauritania is facing low to medium water stress in the region. In addition to water stress, one of the serious difficulties facing the Maghreb authorities are in the regime of water scarcity management. It is currently at a crisis level which needs a quick reaction from all stakeholders.

Instead, with the ever-increasing demand of water to satisfy human development across the world and particularly the Maghreb region, it is observed that the most people in the world could not use this essential natural resource effectively, in terms of use, conserve and repurpose, specifically for the best regional management of water resources. It is important for countries and regions to urgently tackle the critical problems presented by water scarcity, determine the root causes of it and articulate new strategies, techniques and good practices to ensure adequate water supply for the needs of growing population. This chapter aims to address the reasons for increasing water consumption and highlights the several strategic-planning water managements through the solutions, techniques and the best practices which the Maghreb region has adopted to address water scarcity.

2. Reasons for increasing water consumption

The causes of water scarcity in the Maghreb region are both natural and human-induced. The primary reasons for increasing water consumption are population grow-urbanization, climate change agriculture and pollution. In fact, the pace of growth far exceeds the pace of regulations to ensure safety, security and sustainability [4].

2.1 Urban population growth

Urban population growth across the world is one of the most influential phenomena affecting water availability. The population of the Maghreb region has increased rapidly over the course of the 20th century. Its present growth rate is among the highest in the world. **Table 1** shows the population grow in different countries of the

Rank	Country	1960 population	2021 population
1	Algeria	11.06 M	44.62 M
2	Morocco	12.33 M	37.34 M
3	Tunisia	4.18 M	11.93 M
4	Libya	1.45 M	6.96 M
5	Mauritania	850.4 K	4.82 M

Table 1.
Population growth in the Maghreb (5 countries); in 1960 and 2021 [5].

Maghreb (in 1960 and 2021). As shown, the overall population has increased in excess of 3 times, i.e., from 29 million in 1960 to 105.67 million in 2021. Algeria, by far is the most populated, with more than 44 million people (by census of 2021). Mauritania is the least populous country in the subregion and has approximately 4.82 million persons (by census of 2021), as shown in **Table 1**. Please note that migrator movements linked to climate disaster and regional conflict are not included in the census of population. The population growth patterns in the Maghreb show a somewhat similar trend. The world population currently is 7.875B as compared to 3.035B in 1960, an increase by 2.59 times of the population in 1960.

Uncontrolled rise in population in the Maghreb region has inordinately increased the demand for water for domestic, industrial, agricultural and municipal needs; it is certain that population growth will further impact water availability per person. The stressed areas in the Maghreb affected more by water scarcity are, also mainly those, with limited supply of water, high population, and even higher population growth rates.

Algeria—In 1960, the population of Algeria was 11.06 m and increased to 44.62 m people in 2021 (**Table 1**). Algeria is currently growing at a rate of 1.85% per year. Around 71.9% of the citizens in Algeria are living in the larger cities in 2020. This trend of average annual urban growth is 2.6% [6], which this has intensified pressure on water supply in cities since the 1990s. Unfortunately, only 30% of people in Algeria have access to safe water to drink [7].

Morocco—Between 1960 and 2021, the number of persons in Morocco have increased from 12.33 m to 37.34 m people (**Table 1**). Morocco's population is growing at a rate of 1.2% per year. Around 67.8% of Moroccan people are living in urban areas and cities in 2020. This increase of urban population in Morocco was reported at around 2% annually in 2020 [6]. Urban population growth in Morocco has increased the demand on the water supply which leads to water shortages and will continue to present specific challenges in foreseeable future, in order to make water available to everyone.

Tunisia—There were only 4.18 m persons in 1960 for Tunisia and it becomes now 11.93 million persons (2021) (**Table 1**). Tunisia's population growth rate from 2019 to 2020 is 1.06%. Around 71.1% of Tunisians people are living in urban areas in 2020. This urban growth rate in Tunisia was reported at around 1.5% annually [6]. Uncontrolled urban population growth in Tunisia remains the most critical socio-economic problem as more supply of water will be needed to fulfill the needs of the increasing population and this will result in a shortage of water supply.

Libya—In 1960 the number of persons for Libya were around 1.4 million and reached approximately 6.96 million persons in 2021 (**Table 1**). The rate of growth in Libya has been increasing since 2015, in 2020, population growth rate for Libya was 1.4%. Around 88.4% of Libyan people are living in urban areas and cities in 2020. This growing trend of urbanization in Libya was reported at around 1.7% annually [6] and has increasingly affected the water reserves and has caused a depletion in the annual per capita of renewable water.

Mauritania—Between 1960 and 2021 the number of people in Mauritania increased from 850.4 K to 4.82 m people (**Table 1**). The birth rate in Mauritania is almost double the worldwide average, leading to an extreme annual growth rate which is currently around 2.5% per year [5]. This high rate in population growth is the foremost driver of Mauritania's water stress, as measured in units of water per capita. Further, more than half of Mauritanians live in urban areas (53%) (2017). Urban growth rates are increasing at a rate of 4.2% annually. In 2017, more than 53%

of Mauritanians live in urban areas and in 2016 the percentage of people living in urban areas increase to more than 60%. Mauritania features the second most rapid urbanization rate in Africa [8].

2.2 Economic growth

In general, growing economies need growing populations, albeit at different rates. However, if population growth exceeds growth of the economy, the standard of living goes down. Industrial development has had an important role in the economic growth, leading to increased manufacturing, which is a water intensive industry using it for processing, cooling, and disposal of waste products. However, industry not only reduces water reserves but also pollutes the sources that remain and thus reducing citizens access to safe water. An overall economic growth will intensify water demand, straining the local water supply to meet the ever-increasing need of a growing population.

Algeria—Algeria's economy is dominated by its export trade in petroleum and natural gas. This sector accounts for around 94% of export earnings in 2019 [9]. Algeria's gas reserves rank as the tenth biggest in the world [10]. Nevertheless, this economic growth in Algeria contributes to biodiversity loss via higher emissions and greater resource consumption such as freshwater reserves.

Morocco—The crucial economy resources in Morocco are agriculture, phosphate minerals, mining, tourism and fish. Morocco has the largest phosphate reserves in the world and ranks as the third-largest producer of phosphate. Industry and mining contribute about one-third of the annual GDP. The recent economic growth in Morocco requires increase in water demand which has imposed a height pressure over national scare water resources and leads to a decrease in the region's biodiversity, in large part because water resources are poorly managed and water stocks are being polluted. Irrigated areas are also compromised because of water shortages and soil erosion, which would further reduce the GDP of Morocco up to 6.7B US dollars per year, as valued in 2016 [11].

Tunisia—Virtually all sectors of the economy rely on water. Tunisia has a relatively diversified economy such as agriculture, mining, tourism and manufacturing production with services contributing nearly 60% to the GDP [12]. The demand for freshwater in Tunisia is likely to increase while supplies decrease due to economic growth and an increase in human population. Drinking water demand was estimated at around 290 Mcm in 1996 and may even reach 491 Mcm in 2030. However, industrial water demand was estimated at 104 Mcm in 1996 and may rise to 203 Mcm in 2030 [13], exerting further stress on water demand.

Libya—In general, Libyan economy depended mainly on the agricultural sector before the discovery of significant oil reserves in 1959. Libya has one of the highest per capita GDPs in Africa, primarily due the oil and gas sector accounts. Economic growth associated with growing population are the main factors behind the rise in water consumption to satisfy increasing water demands in Libya either in agriculture (the growth of irrigated agricultural areas) or industrial uses. As Libya is already facing a critical water shortage, it is increasingly depleting its precious groundwater resources, most of which are non-renewable.

Mauritania—Mauritania's economy is still largely based on agriculture including crops, livestock and fisheries. At present, fisheries and marine resources account for more than 12.5% of Mauritania's GDP. This country depends also on extractive industries; iron ore, which account for almost 50% of total exports. Additionally, an expanding mining industry (rise in metal prices, gold and copper) and recently found oil reserves led Mauritania's recent economic expansion. Despite the country's huge reserves of resources (fish, iron, oil, gold, etc.), it counts as a lower middle-income country as more than 16.6% of the population lives below the extreme poverty line. The informal economy accounts for 40% of GDP and employs about 86% of the working-age population [8]. Rural areas in Mauritania suffer regular shortages due to the country's periodic droughts in the 1970s, 1980s and 2000s; declining rainfall and the restricted technical and financial capacity of the Mauritanian National Rural Water Supply Agency, often force the migration of farmers to cities and cause more pressure on water reserves.

2.3 Climate change

The Maghreb countries are located in an arid to semiarid region. The climate is Saharian in the southern, oceanic in the western, and Mediterranean in the north. In the Maghreb, rising temperatures associated with climate change during the last half-century reveal a significant level of vulnerability, related, among other things, to:

- Significant degradation of arable land;
- Degradation of pasture and loss of livestock;
- Degradation, even disappearance, of forests;
- High risk of collapse of coastal dune bar;
- Decrease of water resources.

Algeria—Democratic Republic of Algeria is the largest country in Africa. More than 80% of the country consists of the Sahara Desert. Over 34% of the population lives in rural areas and is heavily concentrated in coastal areas. Temperature rises over the last 30 years in Algeria are expected to be further amplified by reductions in rainfall which is estimated at 30% [14]. This rainfall deficit is putting more pressure on the groundwater resources to satisfy water demands and thus a decline in piezometric level of major aquifers in the north part; greater aquifers depletion which reach more than 20 m in certain parts of the region [15].

Morocco—The climate in Morocco is Mediterranean on the coasts, humid temperate at higher elevations and saharan (hot and dry) in the south-western part. Over the last 5 decades, drought events have intensified in Morocco resulting in less rainfall; with possibly, up to at least ~50%, coupled with overexploitation of groundwater resources to satisfy the water demands. The Mikkes basin is located in the North-central of Morocco with a deficit to be estimated around 76% (between the periods 1970–1979 and 1980–2000); this strong deficit cause high depletion in piezometric level (since 1980). This, in turn, has an influence on springs and River flow [16–20].

Tunisia is a country on the Mediterranean coast of North Africa. It situated between Algeria and Libya along the Mediterranean Sea. It covers an area of 164,000 km^2 and over 1300 km of coastline along its eastern and northern borders. Tunisia has a semi-arid to arid climate, it has also two distinctives Mediterranean coasts, west–east in the north, and north–south in the east. Tunisia is considered to be one of the countries most exposed to climate change in the Mediterranean; increasing temperature, less rainfall and rising sea levels. With episodic droughts coupled with floods and the salty types of rocks found within the country all are combined factors which could severely disturb agricultural practices and availability of water reserves (which are already limited). Consequently, several farms lowered their wells and pumped more water to satisfy their needs, which will further put more pressure on already stressed aquifers.

Libya is a country with around 95% of its areas are deserts [21]. The climate of Libya is a Mediterranean climate which is characterized by a cool rainy winter season and a hot dry summer. Thus, this country is susceptible to more frequency of droughts, floods, sandstorms, dust storms, and desertification. Freshwater resources of Libya are very limited and originate mainly from four aquifers—Kufra, Sirt, Morzuk, and Hamada—the last three are close to depletion. In fact, people living in this region are facing absolute scarcity of water (below 500 m^3 per year) and they will face more water risks across the supply with climate variability and its effects (rising in temperature and decline in precipitations). Libya's long periods of drought is expected to cause more threat in crops produce, particularly in areas where water supplies are already under pressure.

Mauritania is a vast Saharan country; it is located in one of the regions most affected by climate change. Mauritania is defined by very scarce and poorly documented water resources', with an average of 2800 m^3/inhabitant/year in 2014 [8]. Surface water (97% of renewable resources), where the majority of the inhabitants live in the coastal cities (Nouakchott and Nouadhi-bou) and in the Senegal River Valley which is used as a primary water source (it is far below the demand for potable water). This River remains at extreme risk of water shortage after being tormented by rising of temperature, drought for several consecutive years and the occurrence of floods and other extreme weather events. Therefore, climate change is certain to accelerate the depletion and degradation of water resources and thus to a reduction in production potential, increased livestock costs, generates more urban agglomerations and reduces grazing lands.

2.4 Agriculture

Urban population growth, increasing temperatures and changing rainfall patterns are among the main factors affecting agriculture production. World population is expected to reach 9.7 billion in 2050 and the global agriculture production has to be increased to meet the increased food demand for this population. As well, human population growth and climate Change have multiple implications on development which shift towards more freshwater withdrawals for agriculture. Similarly, in the Maghreb, continuously growing population will be demanding for more food, more land used for crops and thus more water.

Algeria is facing really scary water shortages because its annual water supply per person is marked less than 1000 m^3 per year. Agriculture in Algeria composes 25% of Algeria's economy. It is a major user of water and is most concerned by the water scarcity. Increased population pressure, economic development, urbanization trends and drought periods are also behind more limited water reserves in Algeria, even in areas where currently rainfall is sufficient.

Morocco—Agriculture in Morocco composes around 15% of Morocco's economy. It accounts for almost 80% of Morocco's water annually [22]. Morocco is facing water scarcity which it is caused by the increase of water demands mainly linked the population growth and climate change. This water risk will likely be more in the near future. According to the World Resources Institute (WRI), Morocco ranked 23rd among countries most at-risk of water shortage. In fact, Innovative irrigation practices can enhance water efficiency. However, they will be too expensive for the average small farmer to afford.

Tunisia is considered as a driest country in the Maghreb region as it has very limited water reserves. It is among the countries which suffers from high water scarcity. Like in Morocco, the agriculture in Tunisia consumes 80% of natural water resources: around 76.9% of the water used for agriculture comes from groundwater and 23.1% comes from surface water [23]. Urban population growth and climate change have combined effect towards a crippling water stress in Tunisia by consuming more water, especially in irrigation sector and this led the depletion rate of Tunisian groundwater resources and also to its vulnerability.

Libya—Like Algeria, Morocco and Tunisia, Libya suffers also from high water scarcity as it has very limited surface water (no permanent River) which composes less than 3% of the total water demands. However, groundwater (both shallow and fossil aquifers), constitute the principal water supply in Libya [24] and it accounts approximately 98% of the total water demands [25]. In 2012, the sector of agriculture used around 85% of the water requirements [26]. In addition, in last over 40 years, the irrigated area has increased and has stressed more water availability due to the lack of rain (droughts) and increased water consumption for irrigation and industries [27].

Mauritania—Agricultural production in Mauritania is primarily subsistence-based and rainfed. Agriculture is thus very localized in the eastern and south-eastern parts of the country, and along the Senegal River. Half the population still depends on farming and raising livestock. Agriculture includes forestry, hunting, and fishing, as well as the cultivation of crops (cereals, most importantly sorghum, in addition to rice, maize, cow peas and millet) and livestock production. Smallholder farmers in Mauritania are increasingly challenged by the uncertainty and variability of weather caused by climate change; recurrent droughts in the 1970s, 1980s and 2000s.

2.5 Pollution

Anthropogenic pollution is a reality in the Maghreb and it is undermining the socioeconomic and ecological basis of life in the region. The fastest growing population, urbanization and effects of climate change are all factors influencing the anthropogenic CO2 and air pollutant emissions across the Maghreb region. Major types of pollution sources in the region include dust storms, sandstorms and gases from various industries. Air pollution and stress on agriculture play a major role in water pollution and realize high number of pollutants (pesticides, herbicides and insecticides) into water bodies and these pollutants not only harm water but also, they treat the biodiversity of the plant and animal life that depend on water to survive.

Algeria—Surface water in Algeria including Tafna, Macta, Cheliff and Seybous River basins are polluted by industrial wastewater and also by organic fertilizers (nitrate, potassium and phosphates) used in agriculture. Groundwater quality in Algeria was generally good. However, this quality has been affected by anthropogenic factors such as pesticides and fertilizers [28]. People in the North part of Algeria consumes more water than others; where the most of its groundwater is non-saline; salinity ≤ 1 (g/l). Nevertheless, certain coastal aquifers in Mitidja and Bas Sebaou are particularly at risk due to intrusion of salty marine water (overexploitation). In the southern Sahara, there are important aquifers, that meet ~96% of water consumption, having salinity ≤9 g/l in the Complex treatment terminal. Additionally, saline lakes with salinity ≥3 g/l are impacted as the quality of surrounding freshwater aquifers and their salinity may increase [29].

Morocco—The increase of water demands in Morocco combined with climate change have led to water deficit which have forced farmers to use wastewater in irrigation and thus affecting the water quality of many water bodies such as streams and groundwater (contamination with a large quantity of nitrogen and phosphates, due to fertilizer runoff from agriculture areas). Also, pressure on groundwater resources in Morocco has been increasing over the last decades which causes a dramatic decline in groundwater levels and degradation in its quality (seawater intrusion, nitrate pollution and natural salinity changes).

Tunisia—Generally, pollution of freshwater water in Tunisia comes from wastewater discharge, industrial waste and agricultural activities. However, Tunisia has the highest access rates to water services and sanitation in the Maghreb region. In the northern coastal region, there is relatively abundant rainfall. In the arid central reason, surface watercourses are ephemeral, flowing only for a few days or weeks a year. Nevertheless, in the South part of Tunisia, surface flows are rare and small. Groundwater is the principal source of water—both shallow renewable and deep, with often non-renewable groundwater resources. The overexploitation of shallow and deep aquifers in this country has led to water-level declines and seawater intrusion.

Libya has very limited freshwater and Libyan people rely heavily on groundwater located underneath the country's vast deserts; about 98% of the total water use [25]. Generally, the water pollution in Libya comes from the combined impact of sewage, oil by products, and industrial effluents. The exploitation of groundwater reserves in Libya has become very crucial in the last decades, especially in coastal areas which compose the only geographic area receiving more than 100 millimeters of rainfall a year and accounts for less than 5% of Libya's land area. In fact, the overexploitation of coastal aquifers reduces freshwater outflow to the sea, and cause a high depletion in piezometric level of these aquifers as much as 25 m below mean sea level which has led to the progressive seawater intrusion in the coastal aquifers since the 1930s [30].

Mauritania lies almost entirely within the Sahara Desert. The only perennial river in the country is the Senegal River, which forms its southern border. The region receives very low rainfall and the only the coastal zone is received significant seasonal rainfall. Several recurrent years of droughts have caused rapidly declined with people settling in rural areas and in shantytowns in cities (where access to clean water and sanitation is scarce). This climate variability put the region in water crisis "water availability" (quantity and quality) and particularly Nouakchott (Capital of Mauritania) at near-constant risk of flooding because it is below sea level and therefore prone to frequent floods caused by rising sea levels and resulting a very water-poor.

3. Impacts of water scarcity in the Maghreb

Increasing water scarcity in response to uncontrolled population growth, industrialization and climate change (rise in food demands and water needs) will further has several impacts on health, habitats and biodiversity.

3.1 Water and health

A lack of adequate sanitation, water quality and malnutrition can be responsible for transmission of many diseases in the Maghreb region which is currently home to more than 105 million individuals. Listed below are the most common diseases in the Maghreb:

1. Diarrhea, one of the leading causes of child mortality in poverty-ridden areas. It is a symptom of a bowel infection which is linked to a lack of safe hygiene practices.

2. Leptospira is frequent in the Maghreb region but is rarely responsible for meningitis.

3. Viral hepatitis represents a serious public health problem in the Maghreb where the prevalence of the 5 viruses A, B, C, D, and E remains high and varies from one Maghreb country to another [31].

4. Typhoid fever (infection caused by the bacterium Salmonella Typhi) is endemic in the Maghreb countries (Morocco, Algeria, Tunisia and Libya). It spreads through contaminated food or water.

5. Schistosomiasis, also known as bilharzia, is a disease caused by parasitic worms that inhabit freshwater rivers and other sources of fresh water. Schistosomiasis is considered one of the neglected tropical diseases (NTDs). It spreads to humans by an intermediate host, namely, freshwater snails. The Maghreb is among the regions with high risk of Schistosomiasis. This disease is caused by lack of hygiene and people get infected during their activities such as swimming in infested water.

6. Leishmaniasis is an infection caused by an intracellular parasite and transmitted to humans by the bite of sandflies. Maghreb is known to be one of the most endemic areas of leishmaniases where both visceral and cutaneous forms are reported [32].

7. Malaria spreads by the bite of female mosquito. With the exception of Mauritania, the countries of the Maghreb have practically eradicated malaria, even though the maintenance phase is underway in Algeria [33].

8. Yellow fever is a mosquito-borne viral disease and is widely spread in Africa. However, there is no risk for yellow fever in the Maghreb countries.

3.2 Impact on habitat

The Maghreb region has limited freshwater reserves. However, there are some permanent rivers that provide fresh water as resource such as Cheliff River in Algeria, Draa and Oum Er-Rbia in Morocco, Latin Bagradas in Tunisia and the Senegal River in Mauritania. As explained earlier the main factors affecting the water reserves in the Maghreb are: exponential population growth—urbanization, climate change, agriculture and pollution. These factors not only cause water risk to human but also threat ecosystems on which both local human populations and aquatics species depend on their survival. For example, when sea level rises as rapidly as it has been; it brings with it many impacts such as (1) wetland flooding in the rainy season, (2) it can cause destructive coastal erosion, (3) the infiltration of sea water in the water-tables which can lead to groundwater quality degradation, (4) the disappearance of low-lying wet lands and all the related biodiversity and (5) lost habitat for fish, birds, and plants. Aquatic ecosystems are the ultimate sinks for the contaminants because of the overuse of pesticides and fertilizers (agricultural activities) and sewage from residential and industrial areas (fecal waste, chemicals, petroleum, sediment). Therefore, water will lose its self-generating capacity and many aquatic species cannot cope with this severe contamination in such ecosystems and also changes associated with the community composition [34]. Knowing that river flow regimes play a key role in freshwater ecosystems. Nevertheless, human activities (industry and agriculture) are known to be severely affecting freshwater ecosystems by rise in temperatures and altering river flow regimes and thereby affecting the habitat conditions and, hence the biodiversity of organisms in surface waters and groundwater.

3.3 Biodiversity degradation

Habitat loss negatively influences biodiversity (which refers to the variety and abundance of different species (animals and plants)) in a particular setting and becomes unable to sustain variety of the species. The Maghreb countries face severe anthropogenic and environmental pressures driven primarily by the water demands of a human population estimated to have exceeded 105 million people in 2021. The Maghreb is an important center of diversity of fauna and flora species which is linked to its great geomorphological variability. Nevertheless, this biodiversity in the Maghreb region is undergoing a serious decline (level of ecosystems, species and populations and genetic diversity). At least six species of native freshwater fish species in the Maghreb region are extinct [35]. The warming up of the sea as a result of climate change affects marine species and ecosystems. For example, some definitively marine species might take the place of certain coastal species. Generally, rising temperatures will endanger the life cycle of some species, cause coral bleaching and the loss of breeding grounds for marine fishes and mammals. Climate change accompanied by the over-pumping has largely depleted groundwater water level in many areas in the Maghreb countries and has led to the deterioration and loss of unique water springs and wetlands with their associated biota. The intensive use of the Maghreb Rivers is projected to increase, exerting immense pressure on the ecosystem and associated impact on the biodiversity; through the disturbance of the biotopes of some species and the dwindling stocks of some populations. Also, the new irrigation schemes may further diminish water supplies in the Maghreb Rivers system and impose additional threats to the biodiversity.

4. Water scarcity management in the Maghreb

Water management is the management of water resources, including ground-, surface- and rain water, to promote efficient use and protect water resources from pollution and over-exploitation. Libya, Morocco, Algeria and Tunisia are currently facing extremely high-water stress (**Figure 2**). As water is a key driver of economic and social development, the challenge of water scarcity for the Maghreb countries has existed for a long time and hence this region is beginning to recognize the importance of an Integrated Water Resources Management (IWRM) which is a holistic approach involving social equity, economic efficiency and environmental sustainability. Consequently, to ensure a reliable, universal and sustainable water supply, there is a need for a transformational water management system to water scarcity in the Maghreb region.

The Regional Initiative on Water Scarcity has been formulated to provide a comprehensive framework to ensure the sustainable use and preservation of scarce water resources in the Maghreb countries. The objectives of The Regional Initiative on Water Scarcity are enhancing policies, investments, governance and best practices to sustainably increase water and land productivity; providing tools for strategic planning of optimal and sustainable allocation of scarce water resources; implementing a regional collaborative strategy for a water-reform agenda. Following proposed initiatives, in addition to the initiatives mentioned above, will further address the water scarcity challenge.

4.1 Preventing water pollution

One of the greatest waters related challenges facing the countries of the Maghreb is the pollution of its freshwater resources which can cause water to become toxic to humans and the environment. A person who wishes to reduce water pollution can help by [36]:

- The right waste in the right bins
- Making sustainable choices regarding food and drinks
- Water conservation
- Disposing of household chemicals properly
- Correct use of fertilizers and pesticides for gardens or farm
- Use of environmentally responsible household products for laundry, household cleaning, and toiletries
- Reducing plastic usage and recycling plastics when possible
- Save the planet and promote awareness about environmental Issues, e.g., growing more plants to reduce drainage of chemicals into the water
- Keeping up with the maintenance of their vehicle to ensure it is not leaking harmful substances

There are several directives legislation adopted in the Maghreb region to stop water pollution including the bathing water directive, monitoring of water and vector-borne diseases and the Drinking Water Directive.

4.1.1 The bathing water directive

The bathing water directive aims to protect human health and facilitating recreational use of natural waters and this through protecting and improving bathing water quality in freshwater and coastal water areas. The "national report on monitoring the quality of bathing water on beaches in the Morocco Kingdom" for 2021 shows that the rate of compliance of the bathing water of Moroccan beaches with microbiological water quality reached 87.06%.

4.1.2 Monitoring of water and vector-borne diseases

Many infectious diseases in the Maghreb region, especially in the aftermath of several disasters, were addressed by installing and functioning of water and sanitation facilities through many legislations which aim to save water and also save lives [37–39]. Moreover, some countries specific directives are listed below:

- Algeria, Morocco and Tunisia have the national communication submitted to UNFCCC includes health implications of climate change mitigation policies and have also conducted a national assessment of climate change impacts, vulnerability and adaptation for health.

- Algeria and Morocco countries have identified a national focal point for climate change in the Ministry of Health.

- Algeria and Tunisia countries have implemented actions to build institutional and technical capacities to work on climate change and health and also have developed Integrated Disease Surveillance and Response (IDSR) system and development of early warning and response systems for climate-sensitive health risks.

- Morocco and Tunisia countries have a national health adaptation strategy approved by relevant government body and have also the national strategy for climate change mitigation includes consideration of the health implications (health risks or co-benefits) of climate change mitigation actions.

- Algeria is currently implementing projects or programmes on health adaptation to climate change.

- Morocco has implemented activities to increase climate resilience of health infrastructure. The National Ministry of Health of Morocco set up basic intervention strategies for the prevention and control of leishmaniases to minimize the incidence of cutaneous leishmaniasis to 50% by 2021 and to avoid mortality related to visceral leishmaniasis [40].

- Mauritania is developed in 2011 the National Sanitation Policy (PNA) and the National Sanitation Strategy (SNA) [8].

4.1.3 The drinking water directive

The Directive in the Maghreb region aims to protect human health from potential dangers and to ensure that drinking water is wholesome and clean. Country specific directives are provided below:

Algeria adopted a monopolistic water management since 1962 to undertake tasks of water stress. The first law instituted between 1962 and 1980 was marked by institutional weaknesses, lack of available water, and nonexistence in demand management capability. Between 1980 and 1999, the second law was manifested by uncertain institutional characteristics and inefficient water management. The third law from 2000 till today has been marked by institutional strengthening, a supply-driven approach, and inefficient water management [41].

Morocco—The main producer of drinkable water in Morocco is the national office of drinkable water (ONEP) that was created in 1972. In 2009, The authorities launched its new National Water Strategy covering the period from 2010 to 2030.This policy is based on the main following strategic objectives: (1) Management of water demand and water efficiency, (2) Management and supply development, (3) Preservation and protection of water resources, (4) Reducing vulnerability to the risks associated with water and adaptation to Climate Change, (5) Modernization of information systems and capacity building and skills and (6) Improvement of the institutional, legal and financial framework [42].

Tunisia—Tunisia's water code was created in 1975; it is the overarching legislation covering the water sector. Despite the scarcity of water resources, Tunisia has embraced a water management initiative which permitted the development of conventional and non-conventional freshwater reserves and control of water consumption in all socio-economic sectors. In 2009, Tunisia water law reform was enacted to reflect the actual social and economic situation in the country. The Water Code remains the most appropriate instrument governing public water domain and any resulting conflicts [43].

Libya—Law No. 7 of 1982 provides protection of the environment from pollution including air, water, soil and food [44].

Mauritania—Law No. 2005-030 of February 2nd, 2005 established the principle of delegation of WSS services by local governments to autonomous professional operators, including private firms. The law also extended the mandate of the Multisector Regulatory Authority (ARE) to the water sector, to regulate service, protect consumer interests and oversee of delegation contracts. However, the implementation of this law suffered from the political and institutional instability between 2005 and 2009, and has to date only marginally replaced GOM's engagement in service operations. Limited progress has been made in transferring service responsibilities from GOM's national operators SNDE or ONSER to private contractors accountable to local governments and ARE [8].

4.1.4 Groundwater projects and regulations

The intensive use of natural resources in the Maghreb, in particular by the agricultural sector has put enormous pressure on freshwater management in different countries of the Maghreb. Groundwater has become one of the most fragile of water reserves to satisfy the rapid development (especially for irrigation) in the region. However, such development has become unstainable because of aquifer overexploitation and its effects (vulnerability). Therefore, the groundwater regulations aim to establish a regime which regulates the environmental quality standards and prevent any inputs pollutants into aquifers, by controlling the direct and indirect discharges of certain substances into its reserves.

Algeria—Water supply (for drinking, agriculture and industry) in Algeria is heavily depend on groundwater. People in the north of Algeria relies on coastal aquifers for irrigation which is facing over-exploitation and vulnerability of its waters. The Algerian government aspires to provide a sustainable water resource to water supply by carrying out a national water plan "The 2005 Water Law". This program is required accurate information on aquifer hydrodynamics to plan and to set up quantitative-qualitative protection areas. The quantitative protection areas are where new wells are banned and the actual abstraction of wells can be limited or discontinued, and the qualitative protection areas are where activities generate pollution [45]. Algeria has also established five different River basin agencies since 1996, after the 1983 legal water was amended [46].

Morocco—The passing of Law 10–95 in 1995 was a major breakthrough in Moroccan water policy. It was a water strategy through River basin Agencies to rationalize water use, provide universal access to the resource, reduce disparities between cities and villages and ensure water security across the country. Also, to control well groundwater depletion caused by overexploitation, the Moroccan authority has used aquifer contracts as a tool in water management [47]. The first Morocco's experience with aquifer contracts began in Souss-Massa region as a technical and non-binding financial assistance by the government.

Tunisia—As mentioned already that only agriculture sector in Tunisia consumes 80% of available water resources. Groundwaters constitute around 43% of irrigation water in Tunisia and mostly are overexploited over the last thirty years [48] which increase the risk of quality degradation in shallow aquifers [49], particularly in the coastal areas. In response to the aquifer degradation and overexploitation, Tunisia has adopted many regulatory and incentive instruments to manage groundwater, such as 1975 Water Law which introduced the concept of the public hydraulic domain, the preeminent act of the state for water management and planning, the protection of ecosystems and the possibility of user-based water scarcity management [43]. Nevertheless, the overall effect has been weak since the water governance still remained weak (regional and local water administrations).

Libya—Great Man-Made River, long-term and enormous project in Libya (2001) aimed to provide Libyan people with their freshwater needs by taking water from reservoirs underneath the Sahara and transporting it along the largest underground network of pipes [50].

Mauritania—Groundwater constitutes only around 3% of water resources in Mauritania. Climate change is likely cause decrease in groundwater recharge and their piezometric levels, especially in the Taoudéni-Tanezrouk and Senegal-Mauritanian basins and their respective aquifer systems. This groundwater vulnerability is getting worsen by anthropogenic pressure due to extraction and deterioration of water quality. In responses to climate change and its effects on water resources, Mauritania has adopted the framework of "the Adaptation Programme of Action on Climate Change (NAPA-RIM)", which aims to identify priority activities that respond to immediate needs to adapt to climate change, ultimately leading to the implementation of national agriculture, livestock and ecosystem protection strategies including:

- Improvement in the monitoring of piezometric groundwater networks and water quality
- Improvement of water resources management
- Establishment of a balance between the availability of water resources and water needs for irrigation and consumption for the population and livestock
- Dissemination of water saving technologies for irrigation

4.2 Best practices and technologies

Globally and more particularly the Maghreb region, changing water availability (quantity and quality) poses complex problems and management options are not easy. As mentioned already, the changing situation comes from different factors including the continuous grow of population and climatic change. The best practices and technologies contributing to water protection in the Maghreb including improved sewage systems, rainwater harvesting and seawater desalination.

4.2.1 Improved sewage systems

Currently, more than half of global population in the world are living in urban areas. In 2050, more than two-thirds of the global population will live in cities [51]. Therefore, economic development accompanied by rapid urbanization increase the use of freshwater resources and may increase also competition for water between cities and agriculture [52] and thereby produce more wastewater. Therefore, it is necessary to treat sewage before disposing it off in a water body as it can cause serious public health concern and also threat aquatic life. Improved and responsible sewage systems along with treatment of wastewater at site and at treatment facilities, are among some of the tools and techniques which help to protect and improve water quality. As such, the component wastewater reuse (for irrigation and other purposes) is emerging as an established water management practice in several water-stressed regions of the world and particularly in the Maghreb region.

Algeria—As stated earlier that water consumption in Algeria is further aggravated due to economic grow, urbanization and climate change which are all main drivers of increasing water demand in irrigation and industrial uses. The reuse of treated wastewater for irrigation (the largest consumer of water) is a priority of the state. In 2005, Algeria allowed use of treated wastewater effluent for irrigation purposes [53] and signed 5 years contract with the SUEZ Company for the management of drinking water supply and sanitation in the greater Algiers area. The contract was renewed for five years in 2011, then for two years in 2016 and finally for three years in 2018.

Morocco—Improved Sewage Systems could contribute considerably to the reduction of 'water stress' and 'water scarcity' in Morocco as part of an Integrated Water Resources Management (IWRM) approach, focusing on the component wastewater reuse for irrigation and other purposes. The development of reclaimed domestic wastewater reuse projects has emerged as a potential non-conventional resource to satisfy the increasing demand for water [54]. On the reuse of treated wastewater, only 12% are currently recycled. This rate increased to 22% in 2020 and may even reach about 100% by 2030 [55]. In 2017, SUEZ are providing Morocco with several wastewater treatment plants and has completed construction and installation of the Anti-Pollution System on the East Coast of Greater Casablanca in Morocco.

Tunisia—Tunisia is a water-scarce country, and water supply security challenges are predicted to be exacerbated by climate change in the coming years [56]. Improved Sewage Systems is a great way that Tunisia practiced for several years as part of an Integrated Water Resources Management (IWRM) approach to satisfy water demands for irrigation and industrial uses. In 2017, the number of operating wastewater plants in Tunisia were 119, producing a volume of dry sludge of 175,000 m^3/y. Tunisia is, therefore, faced with the challenge of finding secure solutions for the recovery and/or disposal of sludge generated from wastewater [57].

Libya—Today Libya is the 20th most water-stressed country in the world. Reuse of treated wastewater was mostly designed in Libya to reduce the water scarcity especially in the irrigation sector. There are around twenty-three wastewater treatment plants distributed all over the country. Nevertheless, only ten are working and in operation [58].

Mauritania—The Mauritanian authorities recently inaugurated a new rainwater collection system as part of a sanitary sewerage network for the city of Nouakchott. This new sanitation network is built by the Chinese company CTE, which aims to enable the evacuation of rainwater and avoid the catastrophic situation that Nouakchott knows whenever there are heavy rains or floods.

4.2.2 Rainwater harvesting

Rainwater harvesting (RWH) system, also called rainwater collection system or rainwater catchment system, and is a likely viable option to increase water productivity at production system level by using rainwater stored in containers such as tanks or cellars. It has been in use since thousands of years almost anywhere and particularly in the Maghreb countries. Rainwater harvesting is an excellent practice of sustainable water management as it helps to meet the growing demand of water and managing scarce rainfall to an extent and also it reduces flood and soil erosion and may decrease drought risk.

Algeria is facing increasingly more serious water shortage problems because the water is unevenly distributed and less available and the rainfall is uncertain and irregular. Algerian are using rainwater collection from the houses roofs as a solution among other seems good in many areas such as Souk Ahras city.

Morocco—The system, developed and installed by Morocco based NGO Dar Si Hmad, is now the world's largest operational fog-water harvesting system which involves around 600 square meters of mesh netting, seven storage reservoirs, six solar panels and more than 10,000 meters of piping. Fog harvesting is an innovative solution to persistent water stress where fog is abundant. It utilizes a specialized mesh to catches the droplets from the fog and gravity pulls the water down into containers, which slowly fills up to conserve a good water reserve [59].

Tunisia—Water harvesting has been practiced successfully in Tunisia and particularly in southeast part to minimize water deficiency. This country encourages new strategy which is promoting water harvesting techniques (surface runoff water harvesting, floodwater harvesting and spreading irrigation) and sustainable farming practices.

Libya faces very severe water scarcity. Rain water harvesting systems is a way for water conservation in the region; its techniques have been increasingly used to get and collect rainwater, storage and prevention dams, cisterns, contour lines and lunar basins.

Mauritania is an arid country in the Maghreb region which is affected by recurrent climate-related shocks, like the drought that threatens agricultural activities and dries up the water table. Rainwater harvesting (RWH) has proved to be a viable alternative water in the Mauritanian capital (Nouakchott) as it enables the evacuation of rainwater and thus fight against floods.

4.2.3 Seawater desalination

Limited natural water resources, the continued population growth and climate change in the Maghreb region are the primary drivers of water stress. Seawater

Desalination is an artificial process of removing dissolved salts from seawater and changing seawater into usable water for human consumption, irrigation, industrial applications, and various other purposes. However, this process produces a highly concentrated brine, which must be properly disposed. This waste product can damage ecosystems if it is not well managed.

Algeria—Desalination of seawater is needed in Algeria to resolve the water issues. Different methods of solar desalination have been practiced such as solar distillation. Combination of reverse osmosis technology and improved water infrastructure has proven to be an effective solution for Algerian cities. Also, the government strategy for drinking water supply is to increase the country's installed desalination capacity. The Tahlyat Myah Magtaa desalination Plant was built in 2011 in Oran city (Algeria's second-largest city) and providing water potable for more than five million [60]. The new seawater desalination plant is being built in Corso, a coastal town 25 km east of the capital Algiers.

Morocco—In order to sustain water needs in Morocco, the government decided to implement a seawater desalination plant in most water-scarce areas (Southern Moroccan) such as Boujdour (MED MVC 250 m^3/d and SWRO 800 m^3/d) and Laayoune (SWRO 7000 m^3/d) [61].

Tunisia—Water is scarce in Tunisia especially in central and southern areas. Water desalination plant may alleviate chronic water scarcity in Tunisia's southern region. The four major desalination plants which produce nearly four percent of the country's total water resource are: Kerkennah (1983) with 3300 m^3/day; Gabes (1995) with 22,510 m^3/day; and two stations in Jerba-Zarzis (1999) with 12,000 m^3/day. Additionally, there are 60 smaller plants in industries services [62]. The plants use reverse osmosis, a process that uses a partially permeable membrane to separate ions, unwanted molecules and larger particles from drinking water.

Libya is a largely desert country which has very limited water resources and suffers from growing water scarcity (urbanization and climate change). Seawater desalination can provide a climate-independent source of drinking water. Desalination technology has been used in Libya since the early 1960s and it is continuously developed during the last 30 years. There are currently 21 operating desalination plants, with a total desalination capacity of 525.680 m^3/d [63].

Mauritania—Water deficits and their associated shortages are serious problems. Thus, the installation of sea water desalination systems is a greatest water technique to help alleviate the shortage of fresh water resources in Mauritania in particular and thus participate in the economic and social development of the Maghreb countries as a whole. In 2018, seawater desalination project designed a 1000 m^3/day in Mauritania's largest fishing port in Nouadhibou for industrial and domestic use [64].

Country	Surface spreading/ infiltration	Open well, shaft and borehole injection	In-channel modification
Algeria	5	—	—
Morocco	1	—	1
Tunisia	6	2	3
Total	12	2	4

Table 2.
Major MAR types in the Maghreb [65].

4.2.4 Recharging aquifers/groundwater

Groundwater is an important source of supply for basic human needs and development across countries of the Maghreb. In many parts of the region, groundwater is the only reliable source of water. As pressures on groundwater resources increase with growing population, economic growth and climate change, there is a need to more practices sources of water supply. Managed aquifer recharge (MAR) is one of these good practices.

Managed aquifer recharge (MAR) also called artificial recharge (AR) is the enhancement of natural groundwater supplies with excess surface water or reclaimed wastewater. It is a process by which excess runoff is directed into the ground accomplished by augmenting the natural infiltration to replenish an aquifer, using man-made conveyances such as infiltration basins, field flooding, infiltration galleries or injection wells [36]. MAR cases are concentrated in highly populated regions of the Maghreb. **Table 2** shows the different MAR cases implanted in the countries of the Maghreb and its different MAR types. The highest number of MAR cases is in Tunisia and lowest one is in Morocco. The most common MAR type in the Maghreb is the surface spreading/infiltration method which is the most practiced in Tunisia (**Table 2**).

Generally, the principal purpose of the MAR schemes is to ensure water supply. The first application of MAR in the Maghreb was started in 1965 (Tunisia) and rise considerably in the 1990s [65].

5. Conclusions

The Maghreb region is facing increasing water scarcity amplified by inefficient water use and overexploitation of water resources (rate grow of population, economic development and climate change considerations). There is evidence that surface water is diminishing and that ground water levels are lowering rapidly. Consequently, variation in water availability either (1) quantitative; the annual volume of water per person in many areas of the Maghreb which is less than 1000 m^3 or (2) qualitative; water pollution has many impacts on Human Health, destruction of habitat and loss of biodiversity. Hence the countries of the Maghreb have tried to overcome water stress and scarcity through: First, preventing water pollution by improving water policy and strategy (monitoring of water and vector-borne diseases). Second, best practices and technologies including improve sewage system, rainwater harvesting, and desalinization, among others. Henceforth, managing water scarcity in the Maghreb region should be proactive rather than being reactive.

Author details

Kaltoum Belhassan
Independent Researcher, Dewsbury, UK

*Address all correspondence to: kbelhassan@yahoo.co.uk

References

[1] MSA. Maghreb Studies Association. The Word Maghreb. 2004. Available from: http://www.maghreb-studies-association.co.uk/en/allhome.html

[2] World Resources Institute. Ranking the World's Most Water-Stressed Countries in 2040. Washington, D.C., United States: World Resources Institute; 2015

[3] Maddocks A, Young RS, Reig P. Ranking the World's Most Water-Stressed Countries in 2040, World Resources Institute, August 26, 2015. 2015. Available from: http://www.wri.org/blog/2015/08/ranking-world%E2%80%99s-most-water-stressed-countries-2040

[4] Vaseashta A, Maftei C, editors. Water Safety, Security and Sustainability. Advanced Sciences and Technologies for Security Applications. Cham: Springer; 2021. DOI: 10.1007/978-3-030-76008-3_19

[5] UN. United Nations, Department of Economic and Social Affairs, Population Division (2019). World Population Prospects 2019, Online Edition. Rev. 1. 2019. Available from: https://worldpopulationreview.com/countries

[6] World Bank. Urban Population Growth (Annual %). 2020. Available from: https://data.worldbank.org/indicator/SP.URB.GROW

[7] Sleet P. Research Analyst, Global Food and Water Crises Research Programme. Water Protests in Algeria Are Giving Cause for Concern about its Long-Term Stability. 2019. Available from: https://www.futuredirections.org.au/publication/water-protests-in-algeria-are-giving-cause-for-concern-about-its-long-term-stability/

[8] World Bank. Mauritania Water and Sanitation Sectoral Project (P167328). Project Information Document (PID). Concept Stage | Date Prepared/Updated: 13-Feb-2019 | Report No: PIDISDSC25021. 2019. Available from: https://documents1.worldbank.org/curated/en/369901553673728547/pdf/Concept-Project-Information-Document-PID-Mauritania-Water-and-Sanitation-Sectoral-Project-P167328.pdf

[9] World Bank. The World Bank in Algeria. 2021. Available from: https://www.worldbank.org/en/country/algeria/overview

[10] Rajakumar A. Petroleum Why The Oil Price Is Falling. Energy & Fuel Users' Journal. Oct.—Dec. 2014. ENFUSE—Volume—LXIII Book- 4. 2014 Available from: http://enfuse.org/wp-content/uploads/2015/05/Oct-Dec2014.pdf

[11] Taheripour F, Tyner WE, Haqiqi I, Sajedinia E. Water scarcity in Morocco: Analysis of Key Water Challenges. Washington, D.C., United States: World Bank Goup; 2020. DOI: 10.1596/33306

[12] Saleh M. Quarterly GDP growth rate in Tunisia 2018-2021. 2021. Available from: https://www.statista.com/statistics/1177784/gdp-growth-rate-in-tunisia/

[13] Benabdallah S. The water resources and water management regimes in Tunisia. In: Agricultural Water Management: Proceedings of a workshop in Tunisia as a Guest. Washington, D.C., United States: National Academies Press; 2007

[14] Touitou M, Al-Aminb AQ. Climate change and water resources in Algeria. Vulnerability, Impact and Adaptation

Strategy. 2018;**18**(1):45. DOI: 10.25167/ees.2018.45.23

[15] Meddiand M, Boucefiane A. Climate change impact on groundwater in Cheliff-ffZahrez basin (Algeria), Elsevier. APCBEE Procedia. 2013;**5**(2013):446-450

[16] Belhassan K. Hydrogeology of the Ribaa-Bittit Springs in the Mikkes Basin (Morocco). International Journal of Water Resources and Environmental Sciences. 2020;**9**(1):07-15

[17] Belhassan K. Relationship between hydrology and climate in the stream Mikkes (Morocco). Research Journal of Earth Sciences. 2011;**3**(1):27-38

[18] Belhassan K. Hydro-geological context of mikkes springs and different variations of their flows (Morocco). Iranian Journal of Earth Sciences. 2010;**2**:143-149

[19] Belhassan K, Hessane MA, Essahlaoui A. Interactions eaux de surface—eaux souterraines: bassin versant de l'Oued Mikkès (Maroc). Hydrological Sciences Journal. 2010;**55**(8):1371-1384

[20] Belhassan K, Hessane MA, Essahlaoui A. Exchange groundwater—river: Stream Mikkes Basin (Morocco). Research Journal of Earth Sciences. 2009;**1**(2):51-61

[21] Makhlouf MH, Etayeb KS. Biodiversity in Libya in Pullaiah, T Global Biodiversity: Volume 3: Selected countries in Africa. Boca Raton, Florida, United States: CRC Press; 2019

[22] Elliese W. An investigation into Morocco's water crisis. WWU Honors Program Senior Projects. 2020;**358**:1-26. Available from: https://cedar.wwu.edu/wwu_honors/358

[23] DRGE (2004) DGRE (Direction Generale des Ressources en Eau). Annuaire Hydrologique de Tunisie 2003-2004. Tunis: DRGE; 2004

[24] Salem OM. Management of shared groundwater basins in Libya. African Water Journal. 2007;**1**:109-120

[25] Shahin M. Hydrology and Water Resources of Africa, Ground-water Resources of Africa. Vol. 2003. New York, USA: Kluwer Academic Publishers; 2003. p. 529

[26] WELL. General Water Authority, 2014, Water and Energy for Life in Libya (WELL). Libya: The European Commission No. 295143, FP7; 2014

[27] Al-Samarrai KA, Sadeg SA. Precision irrigation efficient technologies practices in Libya from the water and energy point of view. International Journal of Applied and Natural Sciences (IJANS). 2018;**9**(6):11-20. ISSN (P): 2319-4014; ISSN (E): 2319-4022

[28] FAOUN. Food and Agriculture Organization of the United Nations, 2009: Draft Synthesis Report. Algeria: Groundwater Management in Algeria; 2009

[29] Safar-Zitoun M. Plan National Secheresse Algerie: Deuxième Version. Paris, France: Convention des Nations Unies de Lutte contre la Désertification; 2018

[30] Alfarrah N, Walraevens K. Groundwater overexploitation and seawater intrusion in coastal areas of arid and semi-arid regions. Water. 2018;**2018**(10):143. DOI: 10.3390/w10020143

[31] Lahlali M, Abid H, Lamine A, Lahmidani N, El Yousfi M, Benajah D, et al. Epidemiology of viral hepatitis

in the Maghreb. La Tunisie Médicale. 2018;**96**(10-11):606-619

[32] Chaara D, Haouas N, Dedet JP, Babba H, Pratlong F. Leishmaniases in Maghreb: An endemic neglected disease. Acta Tropica. 2014;**V**(132):80-93. DOI: 10.1016/j.actatropica.2013.12.018

[33] Ba O, Ouldabdallahi M, Koïta M, Sy O, Dahdi SA. Epidemiology of malaria and elimination prospects in Maghreb countries. Tunisia Medical. 2018;**96**(10-11):590-598

[34] Abou-Elela SI, Zaher F. Pollution prevention in the oil and soap industry: A case study. Water Science and Technology. 1998;**38**(4):139-144

[35] Ford M, Brahimi A, Doadrio I, Perea S, Yahyaoui A, Freyhof J. Freshwater fishes of the Maghreb: Distribution, diversity and conservation status. In: Frontiers in Marine Science Conference Abstract: XVI European Congress of Ichthyology. Bern, Switzerland: European Ichthyological Society; 2019. DOI: 10.3389/conf.fmars.2019.07.00136

[36] Belhassan K. Water scarcity management. In: Vaseashta A, Maftei C, editors. Water Safety, Security and Sustainability. Advanced Sciences and Technologies for Security Applications. Cham: Springer; 2021. DOI: 10.1007/978-3-030-76008-3_19

[37] WHO. World Health Organization & United Nations Framework Convention on Climate Change. In: 2015 Climate and Health Country Profile 2015. Algeria: World Health Organization; 2015. Available from: https://apps.who.int/iris/handle/10665/246137

[38] WHO. World Health Organization & United Nations Framework Convention on Climate Change. In: 2015 Climate and Health Country Profile 2015. Morocco: World Health Organization; 2015. Available from: https://www.who.int/publications/i/item/health-and-climate-change-country-profile-2015-marocco

[39] WHO. World Health Organization & United Nations Framework Convention on Climate Change. In: 2015 Climate and Health Country Profile 2015. Tunisia: World Health Organization; 2015. Available from: https://apps.who.int/iris/bitstream/handle/10665/246121/WHO-FWC-PHE-EPE-15.46-eng.pdf?sequence=1&isAllowed=yhttps://apps.who.int/iris/bitstream/handle/10665/246121/WHO-FWC-PHE-EPE-15.46-eng.pdf?sequence=1&isAllowed=y

[40] Kholoud K, Denis S, Lahouari B, El Hidan MA, Bouhout S. Management of Leishmaniases in the era of climate change in Morocco. International Journal of Environmental Research and Public Health. 2018;**15**:1542. DOI: 10.3390/ijerph15071542

[41] Laoubi K, Yamao M. Water policy reforms in Algeria's agriculture: A review and prospects. Water and Society. 2011;**V**(153):435-446. DOI: 10.2495/WS110381

[42] Benbiba M. Une nouvelle stratégie nationale de l'eau pour une meilleure adaptation aux changements climatiques. In: Proceedings of the International Conference on Water Scarcity and Drought, 18-19 February 2010. Madrid, Spain: Taylor & Francis; 2010 (in French)

[43] Closas A, Molle F. Groundwater Governance in the Arab World—Taking Stock and Addressing the Challenges. Colombo, Sri Lanka: IWMI Project Publication; 2016

[44] FAO. Environmental, Climate and Sustainable Development Laws. 2010. Available from: https://

www.ilo.org/dyn/natlex/natlex4.detail?p_isn=83500&p_lang=en

[45] Algérie. Loi n°05-12 du 28 Joumada Ethania 1426 correspondant au 4 août 2005 relative à l'eau. Journal Officiel de la République Algérienne. 2005;**60**:187

[46] El-Gohary F. Incorporation of Environmental Dimension in Integrated Water Resources Management, Water Governance Program in the Arab States. Cairo: UNDP; 2010

[47] Del Vecchio K, Barone S. Has Morocco's groundwater policy changed? Lessons from the institutional approach. Water Alternatives. 2018;**11**(3):638-662

[48] MA (Ministère de l'Agriculture). Stratégie nationale d'adaptation de l'agriculture tunisienne et des écosystèmes aux changements climatiques. New York: United Nations Development Programme; 2007

[49] Trabelsi R, Zairi M, Smida H, Ben Dhia H. Salinisation des nappes côtières: cas de la nappe nord du Sahel de Sfax, Tunisie. Comptes Rendus Geoscience. 2005;**337**(5):515-524

[50] GMR. Great Man-Made River Water Supply Project. 2001. Available from: https://www.water technology.net/projects/gmr/

[51] Jiang L, O'Neill BC. Global urbanization projections for the shared socioeconomic pathways. Global Environmental Change. 2017;**42**:193-199

[52] Flörke M, Schneider C, McDonald RI. Water competition between cities and agriculture driven by climate change and urban growth. Nature Sustainability. 2018;**1**:51-58

[53] World Bank. World Reuse in the Arab World from Principle to Practice. A Summary of Proceeding Expert Consultation. Wastewater Management in the Arab World.: Dubai-UAE; 2011

[54] Choukr-Allah R, Hamdy A. Wastewater treatment and reuse in Mediterranean region as a potential resource for drought mitigation. Options Méditerranéennes: Série B. 2004;**47**:219-243

[55] Salama Y, Chennaoui M, Sylla A, Mountadar M, Rihani M, Assobhei O. Review of wastewater treatment and reuse in the Morocco: Aspects and perspectives. International Journal of Environment and Pollution Research. 2017;**2**(1):9-25

[56] World Bank. Water and Sanitation for All in Tunisia: A Realistic Objective. Washington, DC : WASH Poverty Diagnostic World Bank; 2018. © World Bank. Available from: https://openknowledge.worldbank.org/handle/10986/30859 License: CC BY 3.0 IGO

[57] Ghacha A, Ammari M, Ben Allal L. Sustainable sewage sludge management in Morocco: Constraints and solutions. Journal of Water and Land Development. 2020;**46**(VII–IX):71-83. DOI: 10.24425/jwld.2020.134199

[58] Brika B. Water resources and desalination in Libya: A review. Proceedings. 2018;**2**(11):586. DOI: 10.3390/proceedings2110586 www.mdpi.com/journal/proceedings

[59] Brittlebank W. UN awards Moroccan Fog-water Harvesting Project. 2016. Available from: https://www.climateaction.org/news/un_awards_moroccan_fog_water_harvesting_project

[60] Hamiche AM, Stambouli AB, Flazi S, Tayeb A. Desalination in Algeria: Photovoltaic power plant for TMM

(Tahlyat Myah Magtaa) of Oran as a case study. In: Negm AM, Bouderbala A, Chenchouni H, Barceló D, editors. Water Resources in Algeria—Part II. The Handbook of Environmental Chemistry. Vol. 98. Cham: Springer; 2020. DOI: 10.1007/698_2020_538

[61] Tahri K. Desalination experience in Morocco. ScienceDirect Desalination. 2001;**136**(1-3):43-48

[62] Holiday L. Agricultural Water Management: Proceedings of a Workshop in Tunisia. Washington, DC: National Research Council, Policy and Global Affairs, Science and Technology for Sustainability Program; 2007

[63] Mukhtar A, Ghurbal SM. Economics of seawater desalination in Libya. Desalination. 2004;**2004**(165):215-218

[64] Xiao J. QT Water Tech Strengthens its Presence in Mauritania. 2020. Available from: https://www.qt-water.ma/en/actualites/9_QT-Water-Tech-renforce-sa-presence-en-Mauritanie

[65] Ebrahim GY, Jonathan FL, Villholth KG. Managed aquifer recharge in Africa: Taking stock and looking forward. Water. 2020;**12**(7):1844. DOI: 10.3390/w12071844

Chapter 10

Water Shortages: Cause of Water Safety in Sub-Saharan Africa

Chelea Matchawe, Patrice Bonny, Germaine Yandang, Huguette Cecile Yangoua Mafo and Bonglaisin J. Nsawir

Abstract

This chapter highlights a high rate of water crisis across sub-Saharan Africa (SSA) despite its huge hydro-potential. Factors contributing to water stress include rainfall deficit and drought, increased water requirements, population growth, urbanization, and poverty. Coupled with the uneven distribution of water resources and mismanagement of water facilities, the gap between the demand for water and available supply has deepened. This has led almost half of the SSA population to drink water from unprotected sources. Moreover, many millions travel far distances and spend several hours daily to collect water. Children and women are mainly involved in water collection. The growing scarcity of water in Africa has a negative impact on economic growth. Besides, water shortages are at the heart of many social crises in SSA and have become directly or indirectly the first cause of death in Africa linked to waterborne diseases. The prevailing water-related diseases include malaria, typhoid fever, cholera, poliomyelitis, etc. To attain the African agenda 2063, national governments in SSA need a multidisciplinary approach integrating, supervising informal settlements of the population in urban and peri-urban areas; improving water storage capacity; increasing irrigation potential for agriculture; and having a good understanding of the epidemiology of waterborne diseases.

Keywords: sub-Saharan Africa, water shortages, negative economic impact, waterborne diseases, contributing factors

1. Introduction

Water is the most basic human need that is required to sustain life on earth. Water is central to plant, animal, and human life. Providing safe water and adequate sanitation is used as a standard indicator to achieve economic development and good health [1]. Therefore, good management of its resources is crucial to the development of every nation. Apart from its vital functions in human life, water is indispensable for the crop, animal, and fish production [2]. However, natural (climate change, rainfall deficit, and drought) and human factors (rapid population growth, urbanization of major cities, agriculture, and tourism) contribute to its scarcity [1, 3, 4]. Water scarcity, defined as lack of sufficient water, or lack of access to safe water supplies, is a global issue [3]. Globally, more than 2.7 billion people face water shortages, and

663 million people in the world lack access to safe water [3]. Unfortunately, half of the people who drink water from unsafe sources live in Africa with 319 million people in sub-Saharan Africa (SSA) [5]. The causes of water shortages may differ in various regions of the world with respect to variations in climatic conditions and socio-cultural realities. Generally, Water shortage can be a limiting factor in poverty alleviation and is associated with negative impacts on the health of the population. It is therefore imperative to understand the driving factors that contribute to water scarcity in SSA and to evaluate its real impacts on the socio-economic development and health of the African sub-continent. Understanding the aforementioned parameters will help to devise appropriate strategies and policies to address the water shortage in SSA. Moreover, a good understanding of how waterborne diseases are affecting the population in both rural and urban areas is necessary to attain the African agenda 2063.

2. The water shortage scenario in Africa

2.1 Water sources in Africa and water shortages: what a contrast

Water scarcity is when water demand exceeds water availability and refers to lack of sufficient water, or lack of access to safe water supplies [3]. There are four dimensions of water scarcity: (i) the first-order water scarcity describes limited available water resources for both current and future needs; (ii) the second-order scarcity highlights lack of financial resources needed to make improved water resources available to the people; (iii) the third-order water scarcity results from the failure of institutional setup or inadequate infrastructure; and (iv) the fourth-order water scarcity is induced by social disparity with the less privileged people experiencing difficulty to have access to safe and sufficient water [6]. These four forms of water scarcity exist in different areas of SSA.

The water crisis is a global issue worldwide. Even though 80% of the Earth's surface is covered by water, fresh water supply has increasingly become a crucial global problem [3]. This global problem received the attention of the United Nations to adopt the Sustainable Development Goal (SDG) target 6 in 2015 [7]. Despite the relative success of the Millennium Development Goals (MDGs) water target in different parts of the world, half of the people who drink water from unsafe sources live in Africa. Specifically, in SSA, 319 million people still live without improved drinking water sources and only 27% of SSA's population has access to basic sanitation [5]. Apparently, Africa seems to be endowed with abundant water resources. The relationship between the abundance of water sources in Africa in general and in sub-Sahara in particular and water scarcity is contrasting. This is why water shortages are experienced even in countries where freshwater is in abundance. Therefore, water scarcity does not mean the absence of water in the natural environment. Actual Africa's hydro potential is irrefutably huge; the continent disposes of 17 major rivers, close to a hundred lakes, coupled with sizeable groundwaters [8]. In addition, Africa has abundant rainfall and relatively low levels of withdrawals of water especially for use in community water supply and agriculture [8]. Inspite of this potential, water sources are unevenly distributed between different zones of the continent [8] with SSA having the greatest part of the water blessing. While the Central Region is endowed with 48% of the continent's water, the Gulf of Guinea claims 24% of the water potential of the continent. At the country level, the Democratic Republic of Congo alone holds 23% of African water [8]. Additionally, natural factors such as rainfall deficit and drought cause significant reductions in the overall availability of water due to the geographical location of several African countries [9]; the present threat on the disappearance of Lake Chad is symptomatic of the growing scarcity of water in SSA. The Lake Chad basin has drastically been reduced by 90% in its surface area [9].

2.2 Factors favoring water shortages in SSA

One of the factors that limit access of the community to safe water in Africa in general and in SSA, in particular, is poverty. The level of poverty is such that half of the extremely poor people live in SSA [10]. It is the main driving force behind the rural migration of the population into the big cities. Due to poverty, many youths back out of school earlier to embrace any money-earning activities in towns. Except, mismanagement and lack of policy prioritization, insufficient infrastructure limiting water supply in SSA has a direct link with the prevailing poverty [11]. Unfortunately 96% of the poorest countries in the world are located in SSA and poverty is projected to increase in this part of the world even in the next 10–12 years [12, 13]. On the other hand, anthropogenic factors including increasing water requirements resulting from irrigation, population growth, and increased urbanization have deepened the gap between the demand for water and available water supply in SSA [4, 14, 15]. The insufficient coverage of potable water in urban areas is particularly attributed to the population growth that has almost doubled over the past 10 years. Consequently, the demand for water supply equally increases. Therefore, the rapid demographic growth together with climate change constitutes a serious challenge for water authorities in SSA. Efforts put in place by the African government often do not yield appreciable outcomes to cope with the ever-growing population. Localities that experience water seizure for over many years rely on alternative sources for water supply. Common alternative water sources used by the population in SSA include wells, boreholes, streams and the rivers. While boreholes can be private or public, streams and rivers are open water sources and are thus owned by local authorities. The correlation between the rapid population growth and the demand for water supply will not be better in the near future as the population is expected to triple by the year 2050 and likely to reach 1.2 billion in SSA [4, 16]. As such, most countries in SSA will be in a state of water stress or scarcity [17]. Moreover, increased demand for water supply can also be induced by the growth of a middle class of citizens in large agglomerations with high water needs. However, the Africa Water Vision 2025 will be based on the principle of service differentiation. The growth of a middle class of citizens with greater water needs will be addressed following this aforementioned principle. Therefore, different socio-economic groups in different parts of an urban area or of a country in Africa will be able to obtain the types and levels of water services that they want and are willing to pay for [18]. Last but not the least, another contributing factor to water shortage in SSA is inadequate water-resources development. Despite a growing demand for water in response to population growth, water scarcity is partly due to low levels of exploitation of water resources. Presently, the entire sub-continent uses less than 2% of its renewable groundwater and irrigates less than 2 MHa (or about 1% of its cultivable land) of groundwater [19]. The current SSA land irrigation capacity is far less than that of the States of Texas in the US. SSA must use modern technologies as other regions such as India do to steer up its agricultural development based on sustainable groundwater development.

Some African towns such as Bamako (Mali), Kampala (Uganda), Lagos (Nigeria), Niamey (Niger), and Ouagadougou (Burkina Faso) will be the most affected by water stress due to their geographical locations coupled with their unprecedented urban growth [7, 20]. Another arising issue that complicates the water crisis in SSA is that the concentration and distribution of formal water outlets is a constant in relation to the consumer demand and the quality of supply that are variables. This phenomenon is most obvious in small towns and peri-urban areas [21].

The stress in the water supply-demand relationship is aggravated by insecurity or socio-political crisis that may cause a shift of a significant influx of the population from crisis areas to other localities. In this case, refugees and internally displaced people constitute the most vulnerable group experiencing water shortages. For example in Cameroon, since 2014 we have witnessed a significant influx of the population from the Far North and, a massive exodus of the population since 2016 from the northwest and southwest to other cities such as Yaoundé, Douala, Bafoussam, etc., because of the insecurity perpetrated by the Islamic sect Boko Haram and the Anglophone crisis. These phenomena have caused the swelling and concentration of populations in certain areas of Yaoundé such as Akok-Ndoe, located in the sub-division of Yaoundé 7. With a population of around 6000 people, this locality does not record any drilling or water supply provided by the public service. Similarly, Mayo-Tsanaga division, located 80 km away from Maroua has been experiencing a significant population growth resulting from the relocation of refugees and displaced persons from Boko Haram attacks. As direct consequences, the population of Mayo-Tsanaga must gather around a single water point to fetch water as seen in **Figure 1**.

On the other hand, the overpopulated area of Akok-Ndoe either dig wells on rocky soil or obtain water from the private water vendors at an exorbitant price ($10 per m^3). Sometimes, the quality of such water is doubtful and is unsafe for the consumers' health. Unfortunately, this situation may depict the reality of many other consumers throughout the sub-continent. The services of the private water providers may compromise the affordability and quality of water, two key criteria for water safety [22].

2.3 Socio-economic impact of water shortages in SSA

There are several dimensions of access to safe water. These comprise proximity, accessibility, reliability, quality, quantity, and affordability [23]. Each of these aspects is almost violated in many countries in SSA. For example, if by standard a water point should preferably be within 200 m, many million across SSA travel for several miles to find a water source. Sometimes at the water site, people wait in a line and carry dirty water into containers to bring home for drinking and cooking. About three-quarters of the households in SSA collect water from a distance far beyond the WHO recommendation [24]. Unfortunately, the burden of water collection and storage usually falls on women and girls [4, 25]. Water fetching is a woman or child-dominated activity in Africa. The impact of distance between the water source and the point of its use goes beyond the physical burden. A study carried out in rural Kenya indicated a relationship between water fetching times and a risk factor for moderate-to-severe diarrhea [26]. In the middle of the year 2020, two children from Tokombere, a sub-divisional headquarter in the Far-North region of Cameroon were found dead on their way to fetch water, due thirst and trekking. Other dangers such as snake and scorpion bites are permanent threats to those that venture to fetch water during hot and cold weather. According to the United Nations estimates, women and young girls spend about 40 billion hours per year transporting water [27]. This corresponds to a

Figure 1.
Women and children struggling to fetch water in the Mayo-Tsanaga Division (Cameroon).

complete year's worth of labor by France's entire workforce [28]. This comprises the time for traveling to the water collection point, waiting at the water source, transporting the water, and storing it.

Even in big cities where water facilities are available, the reliability dimension of water is not always ensured. The high water demand in contrast with the limited water supply reduces the pressure of water flow in the water pipeline (water not flowing in some taps) leading to a lack of water, especially in the morning hours. This may cause water shortage for days or weeks in some countries like Cameroon. Sometimes, a rupture of water tank can lead to a lack of water supply to hundreds of people. Unfortunately, it takes many days for water authorities to be aware and to address the situation.

In addition to their public health impact, waterborne diseases can have a significant impact on the economy of endemic countries [29]. Water appears to have an economic value and should be recognized as an economic good. Providing clean water and a healthy environment is used as a standard indicator of achieved development as highlighted in the SDG water target. Otherwise, water shortage constitutes a serious setback to sustainable development. Water shortage can be a limiting factor in poverty alleviation resulting in low productivity, food insecurity, and constrained economic development [18]. This is because inadequate water resources can restrain improved agricultural development given that agriculture is the largest user of water in Africa. Water stress is particularly a serious threat to irrigated agriculture leading to food insecurity.

The water crisis is at the heart of many social tensions in SSA. Water shortages have plunged several countries in SSA into major social crises. For instance, the crisis in Darfur (Sudan) and recently the tribal conflicts between the Musgum and the Chua Arabs in Logone and Chari (Northern Cameroon) stem in part from water disputes [30] as seen in **Figure 2**. In Darfur, the conflict resulted from competition over water and grazing land between two groups of nomadic farmers, in the same light, in the Logone and Chari division of the Far-North of Cameroon, the Musgum (mainly farmers and fishermen) are competing with the Shua Arabs (herders) over increasing scarcity of water and land resources resulting from the reduction of water bodies in the Lake Chad Basin. Other water-based conflicts in SSA include violent conflict between Senegalese and Mauritanians over the introduction of the irrigation systems, conflicts in the Niger Delta resulting from the struggle over access to limited wetlands due to the decrease of the level of the Niger River, etc. [30].

A B

Figure 2.
A: Displaced Musgum in Kousseri. B: Darfur conflicts in South Sudan.

2.4 Water shortages a threat to water quality and cause of waterborne diseases in Africa

Apart from its negative socio-economic impacts, inadequate access to safe water remains a high risk for communicable diseases that in return reduce vitality and economic productivity. The serious water shortages in SSA has forced communities to rely on unsafe water sources. Unfortunately, the water of these sources is often used without any form of treatment [31]. Almost half of the people drinking water from unprotected sources live in SSA [32]. This explains why more than 70–80% of diseases on the African continent are related to poor water quality [32].

Even in the presence of water availability, water management practices at the level of households have also a great responsibility in spreading water-borne diseases [33]. For instance, lack of awareness, knowledge, and hygiene practices could therefore be barriers to safe water use. Poorly managed sanitation facilities expose water resources to contamination [4]. For example in Cameroon, recurrent ruptures of septic tanks dumping their content on the main roads connecting working-class neighborhoods in cities like Yaoundé are recorded daily [34]. Such human wastes are carried by rainfall and sometimes end their course in a river or any other water source thereby exposing the community to serious waterborne diseases. Other wastes that contribute to polluting water sources in our major cities in SSA include plastic bags and plastic bottles. As they accumulate in the rivers, they can divert the direction of the water flow into the community. This may explain why the quality of the water sources often correlates well with the prevalence of water-related diseases in the community [35]. Additionally, lack of waste treatments in urban areas, insufficient water treatment facilities, as well as mismanagement of the existing water facilities are among the factors that contribute to the deterioration of Africa's water quality [34, 36, 37]. In addition, the use of chemical contaminants in the cultivable areas of our cities represents a source of risk of contamination of water from wells, boreholes, etc. The presence of these products might result to heavy metals in various water sources across SSA and poses as much a public health problem as microorganisms [31, 38, 39]. In addition to the questionable quality of water due to water shortages, urbanization and population growth also contribute to the disposal of more wastes into water bodies in many countries in SSA [38].

Worldwide, the annual loss of human life associated with the consumption of unsafe water is estimated at 30 million people. It thus appears that, for lack of proper access to the resource, water has become directly or indirectly the first cause of death in Africa [8]. Water-related diseases constitute a significant proportion of the burden of disease in SSA.

Diseases resulting from the use of unsafe water or water stress can be grouped into (i) waterborne diseases (e.g. cholera, typhoid, etc.); (ii) water-related diseases (e.g. malaria, yellow fever, river blindness, sleeping sickness, etc.), (iii) water-based diseases (e.g. guinea worm and bilharzia etc.), (iv) water-scarce diseases (trachoma and scabies, etc.).

Diarrheal diseases be of viral, bacterial, or parasitic origin are the leading cause of human mortality in Africa. Our continent alone contributes to 53% of the diarrheal cases reported globally, with contaminated drinking water being the main source of transmission [40]. In 2016, more than half a billion deaths in SSA were attributed to diarrheal diseases with contamination of drinking water identified as one of the leading risk factors. Mortality due to water stress coupled with poor sanitation and hygiene is projected to substantially increase by 1.5 deaths per 1000 annually by the year 2050 [41]. This is true for countries with high mortality rate such as Angola, Burkina Faso, Burundi, Central African Republic, Chad, DRC, Ethiopia, Guinea Bissau, Liberia, Mali, Niger, Sierra Leone, and Somalia [41] (**Figure 3**). The most devastating waterborne diarrheal disease on the African continent is cholera, which is caused by Vibrio cholera. Cholera is a deadly diarrheal disease that decimates tens of thousands of people annually. Approximately, more than one million cholera cases are reported in Africa [42]. This may explain why 83% of the total deaths due to cholera were from the SSA region [43]. Additionally, a curated database of cholera incidence in SSA from 2010 to 2020 identified 999 suspected cholera outbreaks across 25 SSA countries [44]. Most of the major outbreaks of this disease occurred in countries such as Nigeria, Cameroon, the Democratic Republic of Congo, Kenya, Ethiopia, and Sudan [42–44]. Notably, the collective outbreaks in four countries alone (Democratic Republic of the Congo, Ethiopia, Cameroon, and South Sudan) represented 65% of total outbreaks that occurred in the entire SSA [44]. Besides poor sanitation and hygiene, floods have been recognized as one of the major contributing factors of cholera outbreaks in SSA [45]. This occurs when floods hinder supply of or access to safe water sources, thereby introducing *Vibrio cholerae* even to areas that are usually not affected by this pathogen [46]. The coastal regions of SSA remain the focal areas. Curiously, most of these cholera foci are in densely urbanized areas of Africa with limited access to safe water and adequate sanitation [42].

The most frequently reported parasitic waterborne diseases in SSA are malaria (95%), schistosomiasis (44.8%), giardiasis (23.4%), soil-transmitted helminths (23.4%), and amoebiasis (21.3%) [27, 47]. Recently in 2020, Malaria infected more

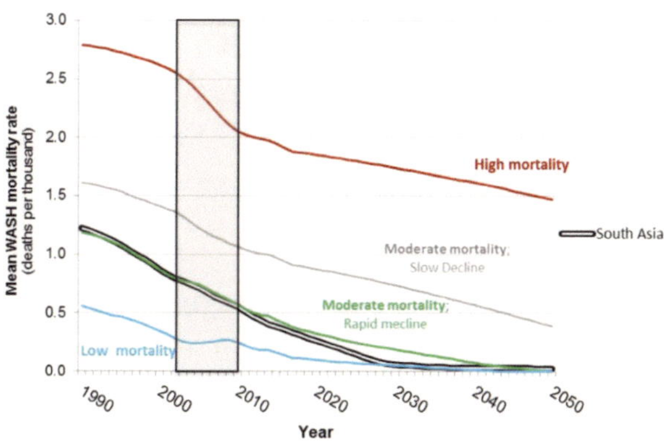

Figure 3.
Projections of WASH mortality rates for SSA countries and South Asia [41].

than 200 million people in SSA indicating its exponential increase attributed to the interruption of malaria-control services during the awake of Covid-19 pandemics [47, 48]. Parasitic infections via water can be acquired while bathing, washing, drinking water, eating food exposed to contaminated water, or being bitten by an infected vector. Interestingly, the prevalence of important parasitic diseases in several regions of SSA is still high in recent years given that most of these diseases circulate in poor water supply areas [29].

Among bacterial waterborne diseases, typhoid fever features as an important cause of morbidity and mortality with an estimated 12–33 million cases leading to 216,000–600,000 deaths annually [49]. Apart from Benin, Equatorial Guinea, Eritrea, Namibia and Somalia, which did not provide any report on typhoid fever, this bacterial disease is highly prevalent over the whole SSA as highlighted in **Figure 4** [50]. The highest incidence of this disease occurs in areas of high water contamination with human feces, limited water supply due to increased population, urbanization, and weak health systems [51]. An updated data on the burden of typhoid fevers from 2010 to 2013 show that this waterborne disease continues to be high in SSA (**Table 1**), and illustrate the need for control measures such as vaccination, and improvements in water quality, sanitation, and hygiene [52]. Gastroenteritis caused by non-typhoidal *Salmonella* is another important waterborne disease that prevails in SSA. Close to 80% of all the reported cases in 2017 occurred in SSA alone, affecting mainly children under 5, adolescents, and active young people under 50 [53].

Many studies across Africa suggest that coliforms, *Escherichia coli*, *Streptococcus*, *Salmonella*, and *Shigella* spp., *Vibrio cholera*, etc. are major contaminants of alternative sources of water in SSA [31, 37, 39, 42].

In the light of its devastating impacts on health and socio-economic developments in Africa in general and in SSA in particular, water crises come immediately after weapons of mass destruction [16]. Consumption of unsafe water thus poses a major challenge to population health in many countries of SSA.

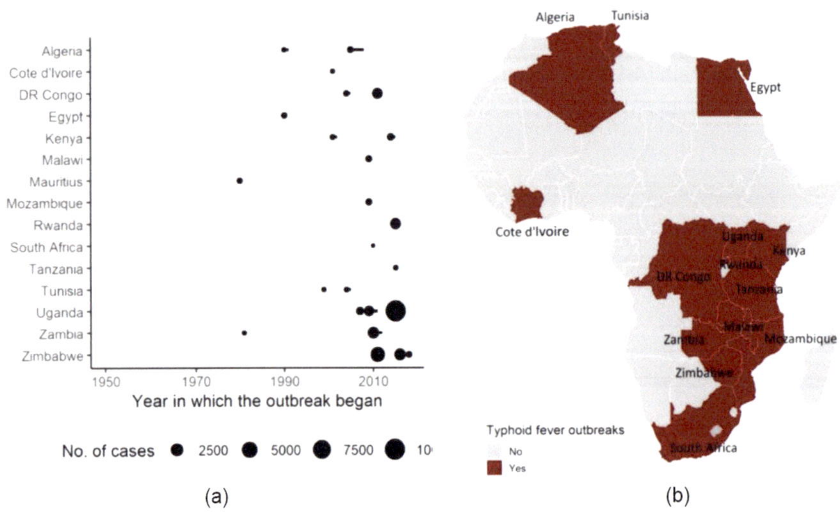

Figure 4.
Prevalence of typhoid fever in sub-Saharan Africa [50].

Published disease burden		Updated disease burden	
	Adjusted*	Adjusted*	Unadjusted
SSA	3056589	3243643	4328528
LMICs	1188304	12070102	20811469

SSA, sub-Saharan Africa; LMICs, low-income and middle-income countries [52].
**Adjusted for water-related risk.*

Table 1.
Typhoid fever burden in SSA estimated from 2010 population data.

3. Conclusions

Addressing water shortage in SSA needs a multidisciplinary approach integrating environmental policy, innovative technologies, and socio-economic dimensions. Preserving and restoring water ecosystems such as wetlands and forests to collect, filter, store, and release water appears vital to reducing water scarcity. Additionally, the reuse of wastewater is another strategy to improve both water availability and quality [3] Implementing a tax on heavy water users such as the industries and agriculture would help avoid wasteful water consumption. Moreover, policies like organic farming practices should be encouraged to reduce water pollution. Improving water storage capacity via the construction of more dams, the storage of water in shallow wells, rainwater collection and storage, drip irrigation for crops is of paramount importance to fight against water shortage. To attain the African agenda 2063 based on inclusive growth and sustainable development, it appears crucial to have a good understanding of how waterborne diseases are affecting the population both in urban and rural communities. Therefore, governments in SSA should supervise informal settlements of the population in urban and peri-urban areas in view to reducing the disease burden resulting from waterborne diseases via improving access to safe water. In this case, future research should be redirected towards environmental determinants of waterborne disease outbreaks, and the relationship between waterborne diseases and water resources development in the context of climate change in SSA.

There is a need for in-depth research with a focus on cross-context and cross-cultural comparisons that can generate important lessons and insights for effective water policies and that take into account different conventional and alternative water uses at different scales. There is also a need for optimal use of groundwater since SSA is currently underusing its renewable groundwater and irrigable cultivable land. National governments should provide adequate investments in water facilities.

Acknowledgements

We register our thanks to AGAPAO, non-governmental organization that gave information about the internally displaced people of Kok-Ndoe in Yaounde 7. We are grateful to the Norwegian Church Aid for the pictures of water crisis in Mayo-Tsanaga and those from conflicts in the Logone and Chari (Far-North Cameroon) and Darfur (South Sudan).

We are equally thankful to the Institute of Medical Research and Medicinal Plants Studies for granting me time for this research.

The corresponding author is also grateful to his darling Irene Assiam for her tremendous encouragement while putting together this chapter.

Conflict of interest

The authors declare no conflicts of interest.

Acronyms and abbreviations

GIZ	Internationale Zusammenarbeit
LMICs	low-income and middle-income countries
MDGs	Millennium Development Goals
PACN	Pan Africa Chemistry Network
SDG	Sustainable Development Goal
SSA	sub-Sahara Africa
UN	United Nations
UN-DESA	United Nations Department of Economic and Social Affairs
UNDP	United Nations Development Program
UNEP	United Nations Environment Program
UNHCR	UN High Commissioner for Refugees
UNICEF	United Nations International Children's Emergency Fund
WASH	Water, sanitation and hygiene
WHO-UN	World Health Organization United Nations

Author details

Chelea Matchawe[1*], Patrice Bonny[1], Germaine Yandang[2],
Huguette Cecile Yangoua Mafo[1] and Bonglaisin J. Nsawir[1]

1 Institute of Medical Research and Medicinal Plants Studies, Yaounde, Cameroon

2 ENSAI, University of Nagoundere, Cameroon

*Address all correspondence to: matchawe@yahoo.com

References

[1] Nwafor CO, Ngene BU, Bamigboye GO, Ogbiye AS, Ogundare JO, Akpan VE. Assessment of water resources development and exploitation in Nigeria: A review of integrated water resources management approach. Heliyon. 2021;7:e05955

[2] Megan H. The Sub-Saharan Water Crisis: An Analysis of its Impact on Public Health in Urban and Rural Nigeria Honor Scholar Theses. 2015:30. Available from: https://scholarship.depauw.edu/studentresearch/30

[3] Dos Santos SE, Adams G, Neville Y, Wada A, de Sherbinin EM, Adamo BB. Urban growth and water access in sub-Saharan Africa: Progress, challenges, and emerging research directions. Science and Environment. 2017;**607-608**:497-508. DOI: 10.1016/j.scitotenv.2017.06.157

[4] He C, Liu Z, Wu J, Pan X, Fang Z, Li J, et al. Future global urban water scarcity and potential solutions. Nature Communications. 2021;**12**:4667. DOI: 10.1038/s41467-021-25026-3

[5] World Health Organization (WHO) & United Nations International Children Emergency Fund (UNICEF). Progress on Sanitation and Drinking Water. Update and MDG Assessment Report. Geneva, Switzerland: WHO & UNICEF; 2015

[6] Chitonge H. Water scarcity for most people in Africa is socially-induced [Internet]. 2020. Available from: https://blogs.lse.ac.uk/africaatlse/2020/12/11/water-scarcity-for-most-people-africa-is-socially-induced-infrastructure/

[7] UN/DESA. World Urbanization Prospects: The 2014 Revision. ST/ESA/SER.A/366 [Internet]. New York. Available from: http://esa.un.org/Unpd/Wup/: United Nations, Population Division; 2015 [Accessed: April 10, 2017]

[8] Dieng M. Water in Africa: The paradoxes of a highly coveted resource. In: Participatory Geographic Information Systems (P-GIS) for Natural Resource Management and Food Security in Africa. 2011. Available from: www.leadinafrica.org/sigp1

[9] UNEP. The tale of a disappearing lake in disaster and conflicts. 2018. Available from: https://reliefweb.int/report/chad/tale-disappearing-lake

[10] World Bank. Poverty and Shared Prosperity. International Bank for Reconstruction and Development. 1818 H Street NW, Washington DC 20433: The World Bank; 2018

[11] Freitas A. Water as a Stress Factor in Sub-Saharan Africa. European Union Institute for Security Studies; 2013. Available from: https://www.iss.europa.eu/sites/default/files/EUISSFiles/Brief_12.pdf

[12] The World Bank. Poverty and shared prosperity 2018: Piecing together the poverty puzzle. Washington, DC: World Bank; 2018. https://www.worldbank.org/en/publication/poverty-and-shared-prosperity-2018

[13] Wadhwa D. The number of extremely poor people continues to rise in sub-Saharan Africa. 2018. Available from: https://blogs.worldbank.org/opendata/number-extremely-poor-people-continues-rise-sub-Saharan-Africa

[14] Beithou N. Non-conventional water resources: Review and developments. International Journal of Research and Science. 2015;**3**:1-9. Available from: www.iaster.com

[15] United Nations. Population Division World Urbanization Prospects: The 2018

Revision (ST/ESA/SER.A/420). New York: Department of Economic and Social Affairs; 2019

[16] Hopewell MR, Graham JP. Trends in access to water supply and sanitation in 31 major sub-Saharan African cities: An analysis of DHS data from 2000 to 2012. BMC Public Health. 2014;**14**:208

[17] PACN. Africa's Water Quality. 2010. Available from: https://www.rsc.org/images/RSC_PACN_Water_Report_tcm18-176914.pdf

[18] Economic Commission for Africa. The Africa Water Vision for 2025: Equitable and Sustainable Use of Water for Socioeconomic Development. Ethiopia: Printed in Addis Ababa; 2015

[19] Cobbing J, Hiller B. Groundwater shortage or crisis narratives are restricting development in Sub-Saharan Africa. Walter Alternatives Forum. 2020. Available from: https://www.water-alternatives.org/index.php/blog/groundwater-africa

[20] Vörösmarty CJ, McIntyre PB, Gessner MO, Dudgeon D, Prusevich A, Green P, et al. Global threats to human water security and river biodiversity. Nature and Resources. 2010;**467**:555-561

[21] Brikké F, Vairavamoorthy K. Managing change to implement integrated urban water management in African cities. Aquatic Procedia. 2016;**6**:3-14

[22] Mvongo VD, Defo C. Assessing water service performances in rural sub-Saharan Africa environment: The case studies of two councils of the southern and eastern regions of the Republic of Cameroon (Central Africa). Journal of Water, Sanitation and Hygiene for Development. 2021;**11**:1. DOI: 10.2166/washdev.2020.164

[23] Internationale Zusammenarbeit (GIZ). Access to Piped Drinking Water. Stresemannstraße, Berlin, Germany: Federal Ministry for Economic Cooperation and Development (BMZ); 2019;**94**:10963

[24] UNICEF and WHO. UNICEF: Collecting Water is Often a Colossal Waste of Time for Women and Girls. 2016. Available from: https://www.unicef.org/press-releases/unicef-collecting-water-often-colossal-waste-time-women-and-girls

[25] Sorenson SB, Morssink C, Campos PA. Safe access to safe water in low income countries: Water fetching in current times. Social Science and Medicine. 2011;**72**:1522-1526

[26] Nygren BJ, O'Reilly CE, Rajasingham A, Omore R, Ombok M, Awuor AO, et al. The relationship between distance to water source and moderate-to-severe diarrhea in the Global Enterics Multi-Center Study in Kenya, 2008-2011. American Journal of Tropical Medicine and Hygiene. 2016;**94**(5):1143-1149

[27] United Nations Development Program (UNDP). Human Development Report 2006. Beyond Scarcity: Power, Poverty and the Global Water Crisis. Houndmills, Basingstoke and New York: Palgrave MacMillan; 2006

[28] Water Project. Poverty and Water. 2021. Available from: https://thewaterproject.org/why-water/poverty

[29] Ngowi HA. Prevalence and pattern of waterborne parasitic infections in eastern Africa: A systematic scoping review. Food and Waterborne Parasitology. 2020;**20**:e00089. DOI: 10.1016/j.fawpar.2020

[30] UNHCR. Dwindling Rains in Northern Cameroon Spark Conflict

and Displacement. 2021. Available from: https://www.unhcr.org/news/stories/2021/11/618ba0ac4/dwindling-rains-northern-cameroon-spark-conflict-displacement.html

[31] Ewang ABA, Chelea M, Bonglaisin JN, Baomog BAM, Adjele JJB, Mouafo HT, et al. The use of alternative water sources as a means of adaptation to water shortages in Nsimeyong, Yaounde city: A quality assessment. Scientific African. 2021;**13**:e00861

[32] Faye C, Dieye S. Valorization of water resources in Senegal for economic, social and sustainable development. African Journal of Environmental Science and Technology. 2018;**12**(12):449-460. DOI: 10.5897/AJEST2018.2563

[33] Wasonga J, Olang'o CO, Kioli F. Improving households knowledge and attitude on water, sanitation, and hygiene practices through school health programme in Nyakach, Kisumu County in Western Kenya. Journal of Anthropology. 2014;**2014**:1-6. DOI: 10.1155/2014/958481

[34] Chelea M, Jideani IA. Survey of gastro-intestinal disease in Cameroon: A case study of the Mayo-Tsanaga Division. Journal of Science and Technology. 2004;**3**(1):1-9

[35] Muller M. Urban water security in Africa: The face of climate and development challenges. Development Southern Africa. 2016;**33**:67-80

[36] Thani TS, Symekher SML, Boga H, Oundo J. Isolation and characterization of *Escherichia coli* pathotypes and factors associated with well and boreholes water contamination in Mombasa County. Pan African Medical Journal. 2016;**23**:12. DOI: 10.11604/pamj.2016.23.12.7755

[37] Kumwenda S, Tsakama M, Kalulu K, Kambala C. Determination of biological, physical and chemical pollutants in Mudi River, Blantyre, Malawi. Journal of Basic and Applied Science and Research. 2012;**2**(7):6833-6839

[38] TheLastWell. What Causes Water Pollution in Africa. 2019. Available from: https://thelastwell.org/2019/06/what-causes-water-pollution-in-africa/

[39] Banda RK, Mubita P, Moonga G, Meki CD. Bacteriological quality and heavy metal analysis of packaged water produced in Lusaka, Zambia and associated quality control measures. Frontiers in Public Health. 2021;**9**:620700. DOI: 10.3389/fpubh.2021.620700

[40] Mutono N, Wright J, Mutembei H, Muema J, Thomas M, Mutunga M, et al. The nexus between improved water supply and waterborne diseases in urban areas in Africa: A scoping review protocol. AAS Open Research. 2020;**3**:12

[41] Fuente D, Allaire M, Jeuland M, Whittington D. Forecasts of mortality and economic losses from poor water and sanitation in sub-Saharan Africa. PLoS One. 2020;**15**(3):e0227611. DOI: 10.1371/journal.pone.0227611

[42] Rebaudet S, Sudre B, Faucher B, Piarroux R. Cholera in Coastal Africa: A systematic review of its heterogeneous environmental determinants. The Journal of Infectious Diseases. 2013;**208**(S1):S98-S106

[43] WHO. WHO Report on Global Surveillance of Epidemic-prone Infectious Diseases. WHO; 2015

[44] Zheng Q, Luquero FJ, Ciglenecki I, Wamala JF, Abubakar A, et al. Cholera outbreaks in sub-Saharan Africa during 2010-2019: A descriptive analysis. medRxiv preprint. 2021. DOI: 10.1101/2021.10.25.21265347

[45] Suhr F, Steinert JI. Epidemiology of foods in sub-Saharan Africa: A systematic review of health outcomes. BMC Public Health. 2022;**22**:268. DOI: 10.1186/s12889-022-12584-4

[46] Rieckmann A, Tamason CC, Gurley ES, Rod NH, Jensen PK, Rieckmann A, et al. Exploring droughts and foods and their association with cholera outbreaks in sub-Saharan Africa: A register-based ecological study from 1990 to 2010. American Journal of Tropical Medicine and Hygiene. 2018;**98**:1269-1274. DOI: 10.4269/ajtmh.17-0778

[47] WHO. WHO Report on Malaria. 2020. Available from: https://www.who.int/news-room/fact-sheets/detail/malaria

[48] Aborode AT, David KB, Uwishema O, Nathaniel AL, Imisioluwa JO, Onigbinde SB, et al. Perspective piece fighting COVID-19 at the expense of malaria in Africa: The consequences and policy options. American Journal of Tropical Medicine and Hygiene. 2020;**104**(1):26-29

[49] Lontuo-Fogang R, Payne VK, Ntangmo TH, Mounchili S, Matango MS, Bup RM, et al. Trends of potential waterborne diseases at different health facilities in Bamboutos Division, West Region, Cameroon: A retrospective appraisal of routine data from 2013 to 2017. Journal of Water and Health. 2021;**19**(4):616. DOI: 10.2166/wh.2021.027

[50] Kim J-H, Im J, Parajulee P, Holm M, Espinoza LMC, Poudyal N, et al. A Systematic review of typhoid fever occurrence in Africa. Clinical Infectious Diseases. 2019;**69**(S6):S492-S498

[51] Andualem G, Abebe T, Kebede N, Gebre-Selassie S, Mihret A, Alemayehu H. A comparative study of Widal test with blood culture in the diagnosis of typhoid fever in febrile patients. BMC Research Notes. 2014;**7**:653

[52] Mogasale V, Mogasale VV, Ramani E, et al. Revisiting typhoid fever surveillance in low and middle-income countries: Lessons from systematic literature review of population based longitudinal studies. BMC Infectious Diseases. 2016;**16**:35

[53] Stanaway JD, Parisi A, Sarkar K, Blacker BF, Reiner JRC, Hay SI. The global burden of non-typhoidal Salmonella invasive disease: A systematic analysis for the Global Burden of Disease Study 2017. Lancet Infectious Disease. 2019;**19**:1312-1324